T0291479

SIR JAMES JEANS

J. H. Jeans.

SIR JAMES JEANS

A BIOGRAPHY

BY

THE LATE

E. A. MILNE

WITH

A MEMOIR

BY

S. C. ROBERTS

CAMBRIDGE

AT THE UNIVERSITY PRESS

1952

CAMBRIDGE UNIVERSITY PRESS
Cambridge, New York, Melbourne, Madrid, Cape Town,
Singapore, São Paulo, Delhi, Mexico City

Cambridge University Press
The Edinburgh Building, Cambridge CB2 8RU, UK

Published in the United States of America by Cambridge University Press, New York

www.cambridge.org
Information on this title: www.cambridge.org/9781107623330

First published 1952
First paperback edition 2013

A catalogue record for this publication is available from the British Library

ISBN 978-1-107-62333-0 Paperback

PUBLISHERS' NOTE

Shortly before his death Professor E. A. Milne had finished the biography of Sir James Jeans which the Syndics of the Press, in consultation with Jeans's executors, had invited him to write, but his failing health had made it impossible for him to give it his final revision. At the request of the Syndics and of Milne's executors, the biographical chapters (I–VI) have been revised by Mr S. C. Roberts, who has also contributed an introductory memoir, and Dr G. J. Whitrow has corrected the later chapters for the Press.

CONTENTS

PLATES

CONTENTS

PLATES

MEMOIR

My first meeting with Jeans was, I think, in 1912 in Charles Sayle's house in Trumpington Street. I was then quite a junior member of the staff of the University Press and Jeans was little more to me than the author of some of those big, blue mathematical books with which I was beginning to be familiar in the Syndics' catalogue. When I returned to the Press after the 1914 war, I began to realize more clearly his importance as an author, but it was not until I became Secretary in 1922 that I had personal dealings with him. Reprints and new editions of his earlier books involved a certain amount of discussion and correspondence, but it was the publication of *Astronomy and Cosmogony* (1928) that led me into more intimate talk with him. I remember very clearly Ralph Fowler coming in to my room at the Press and asking me whether I had read Jeans's latest book. I took the enquiry to be a jocular one and reminded Fowler, in reply, that I was not obliged to read every book that I published. Then, more seriously, Fowler said: 'Ah, yes, but you should look at the last chapter.' It was good advice and I realized, especially after promptings from my colleague, R. J. L. Kingsford, that cosmogony might contain the potentialities of best-selling beyond the dreams of academic avarice.

At that time I frequently travelled by road to Worthing, where my parents lived. Jeans's home at Dorking was only a few yards off the main road and accordingly I proposed myself for lunch on a day when I was due to go to Worthing. It was the first time I had seen Jeans at home and he gave me a most friendly welcome. He produced an admirable claret and after lunch we retired to his study. After a few preliminary *pourparlers*, I approached my main topic and asked Jeans whether he would consider the writing of a popular book. His reply was characteristic. Looking at me with a kindly but slightly scornful expression, he said: 'Oh, yes, several publishers have approached me about that.' 'Well', I replied, 'what about us?' 'Oh', he said 'you're the finest mathematical printers in the world—but you couldn't sell a popular book.' 'Well, have you ever written one?' I countered. From that moment onwards the situation was easier. I could see that Jeans set a definite value on being published by a press famous for its high standard of printing; and, as I afterwards realized, he was also keen to maintain his friendly rivalry with Eddington, whose *Nature of the Physical World* was one of the publishing successes of 1928.

Jeans never haggled over royalties and an agreement was signed in April 1929 for a book to be entitled *The Universe Around Us.* We took a good deal of trouble over the illustrations and the book was published in September. The first edition was one of 7500 copies and was sold out during October. By the end of 1929 11,300 copies had been sold. At the Press we were, of course, delighted and so, in his own way, was Jeans. By that time I had begun to realize that what prevented him from a full display of pleasure or enthusiasm was his shyness. He could rap out a caustic criticism of anyone or anything in a way that chilled his hearers and made them shrink from pursuing the conversation. But when he was genuinely pleased, he found it difficult to express himself. There was no doubt about his satisfaction at the success of *The Universe Around Us,* but it was with an effort that he said to me: 'I always thought the book would do well, but you've sold more copies than I thought you could.' I confess that I felt rather as Boswell did when he had contrived to take Johnson out to dinner to meet John Wilkes: 'I exulted as much as a fortune-hunter who has got an heiress into a post-chaise with him to set out for Gretna-Green.'

In the following year, 1930, Jeans was invited by the Vice-Chancellor (A. B. Ramsay, Master of Magdalene) to deliver the Rede Lecture before the University. As soon as I heard the announcement I wrote to Jeans telling him that we would of course like to publish the lecture and asking him to let us have 'copy' in good time so that we might have the lecture on sale immediately after its delivery. By this time, Jeans was becoming alert to the potentialities of popular publishing and he not only promised to send the manuscript well in advance, but suggested that the book might be considerably longer than the lecture. In this I cordially concurred and a few weeks before the delivery of the Rede Lecture we had 10,000 copies of *The Mysterious Universe* printed and bound. The lecture was to be given in the Senate House at 5 p.m. on 4 November. Two days before this date I was rung up by my old friend Harold Child of *The Times*. 'S.C.', he said, 'the whole office is buzzing about Jeans. Can you let us have the manuscript of the lecture in advance?' I replied that I could, adding a warning that review-copies of the book, as distinct from the lecture, had, of course, been distributed to every newspaper. So *The Times* had its early sight of the Rede Lecture, and, on the morning after its delivery, had a 'turnover' on the middle page, together with a leading article. In the middle of the morning Jeans came into my room. I could see that he was immensely pleased, but he had the embarrassed air of a sixth-form boy who had just won a scholarship and wanted to thank the form-master for his help. 'I've got a very good show this morning', he said, and added jerkily, 'thanks to you, I expect.'

For the next few weeks our chief concern was to keep *The Mysterious Universe* in stock. Walter Lewis, our printer, cleared the decks with great gusto and we sold 1000 copies a day for a month. Reviewers in a wide range of journals wrote about the book at length and country vicars introduced it into their sermons.

By this time I began to feel that I could talk to Jeans as a friend and not merely as a successful author. I paid frequent visits to Cleveland Lodge and saw a little of his daughter Olivia when she came up to Newnham. One popular book followed another and I remember one harassing day which Jeans and I spent in London in an endeavour to settle a dispute between the *Sunday Express* and the B.B.C. *The Stars in Their Courses* (1931) was an expansion of a series of broadcast talks and we had sold the first serial rights to the *Sunday Express*. The first of the Sunday articles happened to correspond rather closely with one of the talks printed in *The Listener*, and the manager of the *Sunday Express* felt that he had a grievance. The legal adviser of the B.B.C. on the other hand, dismissed the claim with contempt. They were both Scotsmen and both obstinate and Jeans and I endured much coming and going between Fleet Street and Broadcasting House before a concordat was reached about 6.30 p.m.

In 1932 my wife died after an intermittent illness of several years; two years later Jeans had to bear a similar blow and the bond between us was strengthened. Quite soon after his wife's death Jeans asked me to spend a night with him at Dorking and for the first time the barriers of shyness seemed to be broken down. He was desperately lonely and unaffectedly glad to welcome me. His daughter was ill in bed and he was clearly pleased to have someone to whom, in some measure at least, he could confide his troubles. Sitting in his lovely garden, he tried to tell me how he felt. 'Sometimes', he said, 'I wish I'd been a games-master at Eton.'

Later in that summer of 1934 Jeans stayed a week-end with me at my house in Barton Road. He was still tired and depressed, but I think he was cheered by the change of scene and company. My mother was also staying with me and I noted with pleased surprise how docilely he listened to her words of old-fashioned advice and comfort.

Early in the summer of 1935 he approached me in his curiously stilted way:

'You never go abroad, do you, Roberts?'

'I haven't been lately', I replied, 'but before my wife's

illness we often used to go to France for ten days or so. Why?'

'Well, I was wondering whether you'd care to come for a holiday with me this year?'

It was the tentative inquiry of a shy and lonely man. I said that I'd certainly like to arrange something—but on a fairly modest scale—and Jeans undertook to obtain particulars from the travel agencies. Towards the end of May I gave a Friday evening lecture at the Royal Institution. Jeans came and talked to me afterwards and we arranged to discuss itineraries on the following morning. After proposing trips to Istanbul and other places, which were quite beyond my resources, he agreed upon Pontresina and to Pontresina we went.

It was my first experience of Switzerland and I entered upon the expedition with a pleasantly ingenuous feeling. What pleased me was that, although I was too old to learn to climb in the technical sense, one could do quite a lot by stout walking. Jeans showed me the Morteratsch glacier in the course of our first day, but his knee was not very reliable and once or twice I went off on my own. One day we went over to St Moritz and spent the day with Helen and Harley Granville-Barker. After an exquisite meal out of doors, described as a picnic, we went over the Maloja Pass and wandered round the picturesque corners of Soglio. Over some very poor beer at the inn, Jeans with the rest of us, became quite hilarious and Granville-Barker proposed that we should send a picture post-card to someone whom all of us knew. We fixed upon Winstanley, then Vice-Master of Trinity and sent him some doggerel beginning:

As duodecimo to folio
Bears very slight affinity
So is the beer of Soglio
To Audit Ale at Trinity....

But Jeans had another expedition to Italy in mind. He persuaded me that it would be a good plan to drive over the

Stelvio Pass and spend a few days at Solda just over the Italian border. He added that he had been asked by his friend Lady Heath to look up a young musician whom he had met at a party in London. At the time I did not realize the significance of this supplementary motive for the journey.

We had an interesting drive over the pass, partly because the car we had hired frequently showed signs of giving out at the steep parts of the ascent. But we arrived at Solda without mishap in the late afternoon. After we had been allotted bedrooms, I came down to the entrance hall of the hotel. Most of the people were Italian tourists, but suddenly a most elegant and individual figure approached—a tall girl in a white climbing-suit. She looked inquiringly at me and we introduced ourselves; she was Susi Hock, the young Viennese musician whom Jeans had met at Lady Heath's party. Then Jeans appeared and our introduction was formalized. Susi had just done a little climb (about 10,000 feet) and went to change for dinner. After dinner we went out on to the hotel terrace. It was a brilliant starlit night and Susi asked many questions about the stars, which Jeans was only too ready to answer. There was also much talk about music. Feigning weariness after a long day, I announced that I was going to bed early. There was no protest. The situation was becoming quite clear to me.

The weather next day was not very good. We climbed up to a nearby hut in the morning. Jeans was a little slow and it was quite easy, and convenient, for me to act as pathfinder. Later in the day Jeans asked me with some embarrassment, whether I would mind if Susi came back with us to Pontresina. I assured him, with complete sincerity, that I should be delighted. The holiday was acquiring an element of romance and excitement for which I had been wholly unprepared. We decided to return to Pontresina by a different route and I quickly announced my desire to sit in front with the driver. On our return to the hotel we aroused some interest. Two days before, we had left—a pair of detached and unromantic widowers. Now we returned with our elegant prize. The dowagers of the Kronenhof were agog with curiosity. But our gaiety was not damped. We went up the Schafberg by the funicular and had lunch at a hut on the way down. There were some odd characters in the hut and we rioted over the silliest jokes. I had never before seen Jeans shaken with helpless laughter.

A day or two afterwards I returned to England. About ten days later I received the following letter from Jeans from Vienna:

'Just a line in haste to tell you—before you see it in *The Times*—that I hope soon to marry Susi Hock.

'I expect this is no surprise to you and fear you must think I owe you an apology for Pontresina and Solda. I am really sorry if you felt it broke up into a 2+1 party, but I had not quite foreseen how things would turn out. Anyhow I hope we may all three meet again soon.'

Thus was the seal set upon my friendship with Jeans. I had a share in the happiest adventure of his later life and although it was always difficult for him to confide frankly in anyone, he would from time to time consult me in a jerky, sceptical way about problems other than those relating to books and publishing. I became godfather to his elder son, Michael, and some years before his death I had agreed to be one of his executors, but I did not know, until his will was read, that he had nominated me as co-guardian, with Susi, of his three children. It was characteristic of him that he had not brought himself to the point of asking me during his lifetime.

Jeans was not a man of many friends, partly because of his temperamental shyness and reticence and partly because of his intolerance of what he deemed to be second-rate. With his own quick perception he lacked the patience which would have enabled him to understand and appreciate a slower-moving mind and consequently he missed those intimacies which he fundamentally desired.

Of Jeans as a man of science it would be impertinent of me to write. As a man of letters handling the problems of the universe he was outstanding and I count myself fortunate to have been not only his publisher, but his friend.

S. C. ROBERTS

January 1952

CHAPTER I

MERCHANT TAYLORS' AND CAMBRIDGE

BORN on 11 September 1877 at Ormskirk, Lancashire, James Hopwood Jeans came of a family of journalists. Both his grandfather and his great-grandfather had owned newspapers and his father's cousin, Sir Alexander Jeans, had been proprietor of the Liverpool *Daily Post* and *Echo*. His father, William Tulloch Jeans, was a parliamentary journalist, representing the *Globe* in the press gallery of the House of Commons. He had a remarkable knowledge of parliamentary procedure and his Fleet Street colleagues always turned to him in their troubles. He was also a keen student of economics and his published works included *The Lives of Electricians* and *Creators of the Age of Steel*.

James's mother, from whom he derived the name Hopwood, came from Stockport and belonged to an evangelical family. Her great-great-great-grandfather had been an Independent minister in Cromwell's time and his small chapel, now used as a school, still stands at Marple, Cheshire. For a time, during James's infancy, his parents lived at Brighton. When he was three years old, they moved to London, living first at Tulse Hill and afterwards at Clapham Park.

James was a precocious child. He could tell the time at the age of three and could read when he was four. He seized upon anything that came his way, even a *Times* leading article which he would read aloud to his parents. The home atmosphere was strictly Victorian, especially in relation to religious observance, and James, naturally shy, began to develop his own interests. He took long walks in London and bicycled into the surrounding country. Later, he accompanied his father very happily on walking tours and the father never ceased to encourage the boy's intellectual development.

From the beginning James displayed a passion for numbers. He could memorize them with ease and at the age of seven made a practice of factorizing cab-numbers. About the same time he came upon his father's book of logarithm tables. He could not make out their purpose, but seized the opportunity of learning the first twenty logarithms by heart. Again, when his mother once lost her ticket on a railway journey, he was able to satisfy the inspector by quoting its number.

Another subject which fascinated him was that of perpetual motion, on which he used to ponder during long services at church. But his greatest enthusiasm was reserved for clocks. All his early drawings contain clocks of all shapes and sizes and sometimes shops full of rows of clocks.

His first written work was a tiny manuscript of nine pages, bound in light blue covers and entitled 'Clocks. By J. Jeans'. The work was fully illustrated inside and out. The text describes the escapement principle in spirited style and gives detailed instructions for constructing a clock out of pieces of tin and other material.

In September 1890 James entered the Merchant Taylors' School as a day-boy. The school, which then occupied the buildings vacated by Charterhouse, was about two miles away and James frequently walked that distance four times a day. Sometimes, however, he went by train and often in company with W.P. (now Sir William) Elderton, with whom he formed a close friendship. Neither of them had a talent for games and Jeans's shyness and his slightly abrupt manner of speaking—characteristics which remained throughout his life—debarred him from making a wide circle of friends at school, but Elderton quickly broke through the shyness and, in particular, derived real enjoyment from Jeans's easy explanations of mathematical difficulties.

Others of Jeans's contemporaries at school were R. V. Laurence, D. A. Winstanley, Evans (of the *Broke*), Cyril Norwood, Herbert Creedy, Major Greenwood, C. H. Reilly,

M. N. Tod, F. J. W. Whipple and F. W. M. Draper. The Headmaster was W. Baker, who founded the Modern Side but preserved the seniority of the Classics. The Classical Side had eight monitors and eight prompters, the Modern only four of each. Jeans reached the lower sixth on the Classical Side in 1893 and then moved over to the Modern Side. At Easter 1894 he was at the top of the upper sixth Modern and two years later he was 2nd Modern monitor and first in mathematics. In November 1895 he won an entrance (major) scholarship at Trinity College, Cambridge.

At Merchant Taylors' Jeans came under three mathematical teachers: the Rev. S. T. H. Saunders, S. O. Roberts, and C. W. Payne. Of these the first became vicar of a city church and at the age of ninety-one could still remember Jeans's quiet industry—'a schoolboy', he said, 'who never got up to mischief'. Roberts, who had been Eighth Wrangler in 1880, was too quick for many of the boys but provided Jeans with a real stimulus towards mathematical study; Payne was a sympathetic and more generally popular teacher. The three of them, as Sir William Elderton remembers, made a good team.

F. W. Morton Palmer (afterwards Scholar of Jesus College, Cambridge, and for many years in medical practice at Worthing) who, under Saunders, sat at the same desk with Jeans recalls that his favourite attitude was to sit with his elbow resting on his crossed knees and his chin cupped in his hand. It was a habit that persisted throughout his life.

W. E. Bowers (later Secretary of the Imperial Continental Gas Company) was, in Jeans's view, one of the ablest of his contemporaries at school. He helped Jeans with his German and Jeans helped him with his mathematics. Many boys, indeed, were impressed by the quickness of Jeans's exposition of a mathematical difficulty. In later years they were astonished at his vogue as a popular writer. At school, they said, he could never see that anything needed explaining. Bowers also remembers Jeans taking a small female part in scenes from *The Critic* at the Speech Day of 1894. What is more important is that he recalls him playing the school organ in the dinner hour. Jeans had, in fact, begun to play the organ at the age of twelve; it was an instrument that was to figure largely in his life.

Jeans went up to Trinity in October 1896 with a Parkin scholarship from his school as well as his entrance (major) scholarship. He read mathematics. His tutor was A. W. Verrall and his director of studies G. T. (later Sir Gilbert) Walker,* who had been Senior Wrangler in 1889.

Other members of the college mathematics staff during Jeans's undergraduate days were J. W. L. Glaisher, W. W. Rouse Ball, A. N. Whitehead, R. A. Herman and E. T. Whittaker.

Men at Trinity reading mathematics who had Rouse Ball for tutor also went to him as director of studies; other mathematical undergraduates went to Whitehead or Walker in alternate years. Those were the days of the unreformed Mathematical Tripos; and Whitehead and Walker thought it unfortunate for a mathematician to spend three years over what was essentially elementary work (Part I of the tripos), and only one year on the living and growing portion of the subject (Part II). They agreed that if they had an exceptionally able pupil they would advise him to risk taking Part I of the tripos in two years, thus leaving two years to work for Part II. In the entrance scholarship examination both Jeans and his Cambridge contemporary, G. H. Hardy, had been outstanding.† Early in the Michaelmas term of 1896 Walker sent for Jeans and Hardy and advised them to take Part I of the Mathematical Tripos

* Director-General of Indian Observatories, 1904–24, and Professor of Meteorology for many years at the Imperial College of Science. Among other studies, Walker is distinguished for his work on the theory of the boomerang.

† Hardy's English essay on 'Historical Novels' might, in the opinion of the classical examiners, have been accepted by the *Nineteenth Century* or the *Fortnightly* for publication; and Jeans's essay was above the average.

in two years. He told them that he could not guarantee that they would come out higher than fifteenth in the list of Wranglers, but he undertook that they would never regret it. They accepted his advice, and went to R. R. Webb, the most famous private coach of the period.

Walker showed courage in giving the advice, since both Jeans and Hardy were potential Senior Wranglers; they, too, showed courage in accepting it. Part I of the tripos had not before been attempted in two years and the risk of a lower place was obvious. The sequel was to justify both Walker and his pupils.

In March 1897 Jeans was elected to an ordinary major scholarship at Trinity and so gained his place on the foundation. At the end of his first year, he told Walker that he had quarrelled with Webb, his coach. Walker accordingly took Jeans himself, and the result was a triumph. In the Part I Mathematical Tripos list of 1898, Jeans was bracketed Second Wrangler with J. F. Cameron (later Master of Gonville and Caius College); R. W. H. T. Hudson was Senior Wrangler and G. H. Hardy Fourth Wrangler. After the results came out, Jeans told Elderton that Hudson was fairly easily head of the field, and that he deserved to be since he could do seven or eight hours of real work a day, and did it, whereas six hours of real work was ample for the rest of them.

Three years of residence were required for a degree, and Part I of the tripos alone, taken in two years, did not carry with it the right to supplicate for a B.A. degree. Consequently, in order to take their degrees in 1899, Jeans and Hardy were obliged to pass the 'special' examination in mathematics and technically they proceeded to ordinary, not to honours, degrees. It may well be that the movement for the abolition of the order of merit in Part I of the tripos was accelerated when it was realized that the best men would from now on take Part I in two years. Be this as it may, the order of merit was abolished in 1909, partly as a result of a pamphleteering campaign in which Hardy took a prominent part.

During his undergraduate career, Jeans had a grand piano in his rooms and took lessons in playing from Kathleen Bruckshaw, the distinguished pianist. He frequently acted as accompanist at undergraduate sing-songs and became secretary of the Query Club.*

His work for Part II of the Mathematical Tripos was interrupted by a tubercular infection of the knee, caused by a minor domestic accident. He left Trinity after the Easter term of 1899, and went first to a sanatorium at Ringwood, Hants, and later to Mundesley, Norfolk, where after two or three years he was completely cured. He returned to Cambridge to take Part II of the Mathematical Tripos in 1900 and was placed in the second division of the first class, Hardy being in the first division. Jeans's lower place was no doubt attributable to his enforced absence from Cambridge. Shortly afterwards he was elected to an Isaac Newton studentship for astronomy and optics, and in 1901 was awarded a Smith's Prize for an essay entitled 'The distribution of molecular energy'. Hardy won a Smith's Prize at the same time, and the two were declared to be Smith's prizemen 'with unspecified relative merit'. The subject of Jeans's essay is of considerable interest as showing that at this early stage he was preoccupied with the great problem of the equipartition of energy in a dynamical system specified by a large number of co-ordinates. A paper with the same title, presumably a version of the Smith's Prize Essay itself, was published in the *Philosophical Transactions* of the Royal Society in 1901.

Jeans was elected a Fellow of Trinity College, Cambridge, in October 1901, and his old schoolfellow, Reginald Vere Laurence, the historian, was elected to a fellowship at the

* One of the Rothschilds, who was a member of the club, was successfully coached for Little-Go by Jeans and marked his gratitude by the gift of a tie-pin, set with diamonds in the form of a query mark. Jeans's sister still possesses it.

same time. In a letter to another schoolfellow, F. W. Morton Palmer, Jeans wrote: 'I am very glad that Laurence has been elected on the same day as myself after the long time we have been together.' Hardy had been elected to a prize fellowship in the previous year, and used to say that he considered his election to a Trinity prize fellowship as the most important event in his career. For Jeans it was similarly of supreme importance since it enabled him without financial worry to pursue the cure of his knee at the sanatoria. The disease slightly affected both wrists and knees before it was finally overcome, in 1902 or 1903.

Jeans took his M.A. by proxy in 1903, and in the following year was appointed University Lecturer in Mathematics. This post he held until he left for Princeton in 1905.

During his enforced leisure at the sanatoria, he wrote his first treatise, the *Dynamical Theory of Gases* (published 1904, second edition 1916), which in its successive editions has been used by generations of students and research workers. The contents of the volume will be described in more detail later.

During his sojourn at Mundesley, Jeans kept in frequent touch with many of the Fellows of Trinity. The following letter (undated) from G. H. Hardy not only shows the pleasant relationship which existed between them, but reveals a certain interest on Hardy's part in 'applied' mathematics, an interest which Hardy consistently disclaimed. The paper of E. T. Whittaker's to which he alludes is presumably that on a general solution of Laplace's equation (*Mathematische Annalen* (1902), **57**, 333).

Trinity College, Cambridge

My dear Jeans,

I was very glad to hear such an encouraging report; I suppose we may really expect you up next term anyhow. Probably you have heard all the news from here, about Sedgwick, etc. It was a dreadful question to vote on: I went against ultimately. The case for was so transparent though that it was very hard to be reasonable.

I was really writing to ask for a copy of your latest paper, which seems to be rivalling Whittaker's in notoriety.

Yours affectionately,

G. H. HARDY

Or, again, this from H. F. Stewart:

30 *Thompson's Lane, Cambridge*
18 Feb. 1903

My dear Jeans,

I am so very glad to hear such a satisfactory account of yourself from yourself, and sincerely trust that you will continue to do well and soon be back here. *Soyez tranquille* on the score of the bedmaker on Staircase I, Nevile's Court. She is a treasure and will look after you with every possible attention.

With regard to a harpsichord, I know of no place where they are stocked! But Mr Arnold Dolmetsch whose address it will be easy to discover could doubtless put your friend into the way of acquiring one. Those that I have knowledge of have all been picked up in farm-houses here and there and have been entrusted to Dolmetsch's care for repairs and tuning.

As for Trinity news I don't know that there is much.—Does the arrival of my infant daughter interest you? She is a fortnight old and has already been vaccinated. She has good long fingers for the piano and even now seems to be wanting to play something. I am looking forward to introducing you to my wife and her when you come up again.

It will be nice to have you underneath me in college. I spend a good deal of time in my rooms and shall descend on you when I want to make music on the spot, as my piano is here.

Forsyth is probably going into G. T. Walker's rooms which leaves a good set but a cold set vacant just opposite. I wonder who will be your *vis-à-vis*. Laurence has transferred himself to Whewell's Court where he is installed with some dignity chez Langley. We are all agog about the new professor [Lucasian]. I wonder who it will be. Personally and apart from mathematics I should like to see Horace Lamb in the post—Himself and his family would be a notable acquisition to Cambridge sorority, and I believe he is worthy of the professorship, is he not?

Now I must stop. When I find Dolmetsch's address I'll write again.

Yours very sincerely,
H. F. STEWART

Finally there is one from R. St J. Parry:

Trinity College, Cambridge
4 May 1902

My dear Jeans,

I had quite hoped to get a mid-week letter off to you this time; but the fates forbade; and here is Sunday again—with an excellent sermon from H. F. Stewart. It was splendid of you to write me such a screed. I wanted to get a kind of idea of what was going on and your letter gave it. I hope too you liked Miss Todd and Miss Jackson—tho' I don't know which of the latter came.

My youngest brother has arrived this week from Persia after five years' absence—very delightful. He has grown a beard; and the bottled-up conversation of five years is being gradually unladen! Also we hear today that my S. African farmer brother is coming home for an operation on his arm. It will be very remarkable that all my brothers will be in England together this summer—7 of us alive. We are having it quite cold; but healthy. Rix has been up this week for the final part of his 3rd M.B.—as delightful as ever but I am afraid overworking. He is very much afraid that he has been ploughed: the result comes out tomorrow.

Gaye and Hardy are taking Jackson and me to the Theatre on Wednesday—to see the *Yeomen of the Guard* or some such! We are greatly excited.

I hope you have enjoyed reading the proposed Statutes. We are sanguine enough to hope that we may get thro' them in one afternoon. I don't know why people should want to talk much more.

I hear from Walker that you are going to take the rooms above his. I believe they are really the best for you among those that are vacant; though I wish you could have come to M.N.C.

Would you care for a copy of the *Essays of Elia* to read again? I have got two copies: they are just the thing for occasional dippings.

I shall not send this if I don't stop now.

Yours ever affectionately,

R. St J. P.

It was in 1905 that Jeans published his definitive solution of the problem of the partition of energy between matter and radiation according to the classical mechanics, and so, by showing that his solution was in rank contradiction with experience, made the acceptance of Planck's quantum theory ultimately certain. The problem had been attacked by Lord Rayleigh, who had indeed found the form of the formula for the theoretical spectrum of black-body radiation (or complete radiation, as it is better called), but Rayleigh gave a wrong numerical factor. Jeans's re-derivation of the formula and his correction to Rayleigh's formula were at once admitted by Rayleigh; and the formula, $8\pi RT\lambda^{-4}\,d\lambda$ for the energy lying between wave-lengths λ and $\lambda + d\lambda$ in complete radiation of temperature T, is known as the Rayleigh-Jeans formula. It will be explained more fully in Chapter IX.

CHAPTER II

PRINCETON, 1905–9

JEANS returned to Cambridge in 1903, completely cured, and settled down to the tenure of his fellowship at Trinity. His earliest research, as we have seen, was concerned with the distribution of velocity amongst the molecules of a gas. Whilst still convalescent, he had engaged in researches, inspired by the work of Sir George Darwin, on the stability of a spiral nebula, and the problem of the forms of equilibrium of rotating and gravitating fluid masses. Jeans had indeed always two strands of interest in mathematical physics, the analysis of the very small (molecular physics) and the analysis of the very large (cosmogony). His actual results in these fields will be described later. Here we note that, though on his return from convalescence he devoted himself principally to molecular physics in the form of statistical mechanics, his work in the two distinct fields had made him known to the leading English and Continental mathematicians of the day. Consequently when, in the autumn of 1904, he was a candidate for the vacant chair of mathematics in the University of Aberdeen, he was able to support his application by testimonials from men of world-wide reputation.

'Ever since we elected him to an open Major Scholarship here,' wrote the Master of Trinity (H. M. Butler), 'his career has been one of exceptional distinction, and he is regarded as one of the most brilliant and original of our younger mathematicians. His character is of the highest, his manners are most agreeable, and he has won the warm regard of us all.'

G. H. (later Sir George) Darwin wrote:

I did not see anything of Mr Jeans during his career as an undergraduate at Trinity College, but a very remarkable essay by him was submitted to me as one of the examiners for the

Fellowship at Trinity College. The paper showed much originality and a very unusual power of dealing with a mathematical problem of great difficulty. The subject was especially interesting to me, and we had several long discussions of it after the fellowship examination. I ultimately had the pleasure of presenting the paper to the Royal Society for publication in the *Philosophical Transactions*. At a later date I presented two other papers by him to the Royal Society, and I consider that the investigations contained in these papers form important contributions to science.

And J. J. Thomson:

I have read many of Mr Jeans's papers, especially those dealing with Electrical subjects, and with the Kinetic Theory of Gases, and have been greatly struck by the originality and power displayed in them. I know of no one among the younger mathematicians who has done abler or more important work in the application of Mathematics to the explanation of Natural Phenomena. I think that any University would be fortunate if they could secure the services of Mr Jeans, whose papers have shown him to be a mathematician of the very highest order.

But Jeans did not go to Aberdeen, H. M. Macdonald being elected to the vacant chair.

It was not long, however, before another chair was offered to him. In the summer of 1905 he arrived at Cape Town for the meeting of the British Association. He was acting as secretary to Section A (physics and mathematics) for the Association, but shortly after landing he received a cable offering him the Professorship of Applied Mathematics at Princeton, N.J. He had no hesitation in accepting the chair; he abandoned the British Association and came straight back to England to prepare for his move.

Jeans was invited to Princeton by Woodrow Wilson, who was then President of the University and was actively engaged in enlarging the faculty at Princeton. The new men in mathematics and physical science were selected by his lifelong friend and adviser, Henry B. Fine, who had an extraordinary flair for picking first-rate young men. The group who came to Princeton in 1905 included Jeans and O. W. Richardson (later Sir Owen) in physics, George Birkhoff, Gilbert Bliss, L. P. Eisenhart and Oswald Veblen in mathematics, and Henry Norris Russell in astronomy.*

Jeans arrived at Princeton with a well-earned reputation. He lectured mainly to the abler students, and to them he delivered advanced courses. Though there had been good teaching in mathematics at Princeton earlier, the intensive type of work in the Cambridge tripos tradition came as something of a shock to the Princeton undergraduates.

It was an old Princeton custom to gather in spring evenings on the steps of Nassau Hall for the singing of college songs. Thus in 1919 the young men sang:

> Here's to Woodrow Wilson. He
> Has left these scenes for Gay Paree.
> We wonder when he'll cease to roam
> And if he'll bring the bacon home.

Not long after his arrival in Princeton, Jeans was elected a Fellow of the Royal Society at the early age of twenty-eight and in 1907 the students were singing:

> Here's to Jimmy Hopwood Jeans.
> He tries to make us Math-machines.
> A young and brilliant F.R.S.,
> That's going some, we all confess.

More important at this time was Jeans's engagement and marriage. He married in 1907 Charlotte Tiffany Mitchell, daughter of Alfred Mitchell, explorer and traveller, of New London, Connecticut. She was connected with the well-known Tiffany family of New York and was a woman of quiet charm, with a genuine gift of poetic feeling and expression.† Her later years were saddened by increasing

* The author is indebted to Professor Russell for much information about Jeans's period at Princeton.

† A volume of her poems, entitled *Driftweed*, with a preface by Jeans, was printed by the Cambridge University Press, for private circulation in 1935.

deafness and a certain melancholy, but the marriage was an extremely happy one. Their only child, Olivia, who was born in 1912, died in 1951.

Jeans remained in Princeton until 1909 and then returned to England for good. Two books belong to his Princeton period: *Theoretical Mechanics* (1906) was a useful and well-planned text-book on statics and dynamics; the other, *Mathematical Theory of Electricity and Magnetism* (1908), was a book of greater importance. Published by the Cambridge University Press in their standard series of mathematical treatises, it went through many editions, and has been used by generations of undergraduates. Its scope is best described in the words of the Preface to the first edition:

> There is a certain well-defined range in Electromagnetic Theory which every student of physics may be expected to have covered, with more or less thoroughness, before proceeding to the study of special branches or developments of the subject. The present book is intended to give the mathematical theory of this range of electromagnetism, together with the mathematical analysis required in its treatment.
>
> The range is very approximately that of Maxwell's original Treatise, but the present book is in many respects more elementary than that of Maxwell. Maxwell's Treatise was written for the fully equipped mathematician: the present book is written more especially for the student and for the physicist of limited mathematical attainments....

The second edition (1911) is arranged in three parts, 'Electrostatics and Current Electricity', 'Magnetism' and 'Electromagnetism'. As Jeans says, the first of these divisions may be said to constitute a development of the anciently known property of amber, that it attracts light objects when rubbed, the second (which should properly be called *magnetostatics*) is a development of the property of the lodestone, and the two are independent of one another. They become related in the third branch, electromagnetism, as developed by Faraday and given mathematical expression by Maxwell. Faraday showed also the connexion between electricity and chemical forces, and part of the connexion between electromagnetism and light, the full connexion being developed by Maxwell.

Jeans began therefore with chapters on the electrostatic field of force, the properties of conductors and dielectrics, and the general theorems on the potential which derive from Green's theorem. He followed this by a long and masterly chapter on special problems in electrostatics, wherein he developed *ab initio* the theory of potential functions, Legendre functions and spherical harmonics, confocal co-ordinates, the theory of images and Kelvin's theory of inversion, and, for two-dimensional problems, the use of the complex variable, conjugate functions and conformal transformations. The chapter is a mine of mathematical information and, though making no pretensions to rigour in the matter of mathematical proof, it has remained a standard exposition of the methods of mathematical physics as applied to electrostatic problems. In spite of its length it is nowhere dull—Jeans never wrote a dull page of mathematics in his life. It is followed by a collection of about a hundred examples of tripos type, many of them very difficult for inexperienced students. This section concludes with two chapters on current flow in linear circuits and in sheets.

The second part consists of two chapters on magnetism, wherein Stokes's theorem is proved and the magnetic vector-potential introduced, and the properties of magnetic shells derived. These chapters are not as successful as the preceding ones, and have roused many difficulties in the mind of the careful reader. The difficulties were, and are, inherent in the subject, rather than of Jeans's making; and to this day there is not altogether agreement about the respective meanings of magnetic induction and magnetic force inside a magnetized body.

The third part deals with the mathematical expression of Faraday's laws of electromagnetic induction, and develops the dynamical theory of currents as a case of general dynamical theory as embodied in Lagrange's equations. It derives Maxwell's equations from the hypothesis of the displacement current, establishes Poynting's theorem, and has a little to say about electrons and electromagnetic waves. But here Jeans is somewhat infringing the limits he set himself, and the reader at this point requires more specialized treatises.

The book is unnecessarily lengthy owing to its failure to use vectors and vector methods. But, by and large, it is a great work, especially in its exposition of the earlier classical portions of the subject. The compilation of its sets of examples must alone have cost great labour. It is written in an easy style which perpetually persuades the reader to go a little further and the continuous demand for new editions is fully justified.

CHAPTER III

RETURN TO ENGLAND

THE ADAMS PRIZE ESSAY

1909–19

SHORTLY after his return to England, Jeans was appointed Stokes Lecturer in Applied Mathematics in the University of Cambridge (1910). He retained this appointment till 1912, when he retired and went to live at Guildford, there devoting himself to mathematical research.

In 1914 Jeans published his justly famous *Report on Radiation and the Quantum Theory* for the Physical Society of London. As a result of the outbreak of war, it did not at first reach a large circle of readers, but it was eagerly read when students of mathematics and physics returned from the war in 1919; and it did much to establish confidence in the quantum theory and in Bohr's then entirely unorthodox theory of the atom and atomic spectra. Together with Eddington's *Report on the Relativity Theory of Gravitation* (1918), also made for the Physical Society of London, it decisively influenced the acceptance by responsible scientists of a new theory.

Jeans's *Report* consisted of seven chapters, of which the first was entitled 'Introductory: On the Need for a Quantum Theory', and the last 'On the Physical Basis of the Quantum Theory'. Of the other five chapters the first two summarized the substantial parts of Jeans's own researches in radiation according to the classical mechanics and the revolutionary modifications of that theory at the hands of Planck, and the remaining three dealt with Bohr's theory of the hydrogen and hydrogen-like atoms and their spectra, Einstein's theory of the photo-electric effect, and the theory of the specific heats of solids, due to Einstein,

Debye and Lindemann. The whole *Report* amounts to ninety pages.

This was of course well before the days of quantum mechanics, and the theory was a collection of dynamical contradictions. At the end of his *Report* Jeans wrote:

> ...It may be asserted with confidence that until some kind of reconciliation can be effected between the demands of the quantum theory and those of the undulatory theory of light, the physical interpretation of the quantum theory is likely to remain in a very unsatisfactory state...the explanation of the black-body spectrum demands the quantum theory and nothing but the quantum theory, all the discontinuities of the theory and their surprising physical consequences included. The keynote of the old mechanics was continuity, *natura non facit saltus*. The keynote of the new mechanics is discontinuity; in Poincaré's words,
>
> > 'Un système physique n'est susceptible que d'un nombre fini d'états distincts; il saute d'un de ces états à l'autre sans passer par une série continue d'états intermédiaires.'
>
> The antithesis is obvious; its resolution will not be easy. Perhaps the present report cannot end better than by a free translation of Poincaré's concluding remarks in his striking article, 'L'hypothèse des Quanta':
>
> > 'We see now how this question stands. The old theories which seemed until recently able to account for all known phenomena have suddenly met with an unexpected check. Some modification has been seen to be necessary. A hypothesis has been suggested by M. Planck, but so strange a hypothesis that every possible means was sought for escaping it. The search has revealed no escape so far, although the new theory bristles with difficulties, many of which are real and not simple illusions caused by the inertia of our minds, which resent change.
> >
> > 'It is impossible at present to predict the final issue. Will some entirely different solution be found? Or will the advocates of the new theory succeed in removing the obstacles which prevent us from accepting it without reserve? Is

discontinuity destined to reign over the physical universe, and will its triumph be final? Or will it finally be recognized that this discontinuity is only apparent, and a disguise for a series of continuous processes?...'

It is of some interest that Jeans should have found no better way of concluding his *Report* than by quoting Henri Poincaré. For there is a profound similarity between the mathematical tastes and activities of Jeans and Poincaré. Both had the gift of mathematical fluency, and both were writers of standard treatises in mathematical physics of great attractiveness and beauty of style. Both were attracted by the world-wide vistas opened up by their studies in cosmogony, yet both made significant contributions to our understanding of molecular phenomena: they were both attracted at once by the phenomena of the very large and the phenomena of the very small. Both made classical contributions to the theory of the forms of equilibrium of rotating gravitating fluid masses and their stability. Both were interested in the philosophical significance of this work. Lastly, both were exceedingly successful as writers of popular works on science: both could expound abstruse physical and mathematical theories in winningly attractive style. As mathematicians, as mathematical physicists, as men of science and as expositors of that science, both were in the highest class. And both sought the philosophical meanings of what they discovered by their researches and of what their results implied.

We have noticed (Chapter I) that the year 1905 saw the publication of Jeans's form of the Rayleigh-Jeans partition formula. Jeans made attempts in the years between 1905 and 1914 to introduce hypotheses which would avoid the 'ultraviolet catastrophe' predicted by the formula, that is, the consequence that in an isolation chamber in a steady state at temperature T, radiation in contiguity with absorbing and emitting matter would perpetually cause the redistribution of energy in the sense of its being sucked into smaller and ever smaller wave-lengths, with no limit to the process. His *Report* of 1914 put the coping-stone to these researches, by finally demonstrating that there was no hope of avoiding the classically predicted ultraviolet catastrophe except by introducing some discontinuous process of the kind suggested by Planck. The year 1914 marked in fact a climax to Jeans's researches in this subject for after 1914 he became more interested in astronomical and cosmological problems, and he never returned to atomic researches. That is to say, he never himself again engaged in research on atomic phenomena or statistical mechanics,* though he was of course to expound them again in his popular works.

Thus 1914 closed an epoch in Jeans's scientific life. This makes his final quotation from Poincaré's *Dernières Pensées* the more appropriate. But it was by no means the end of Jeans's active participation in research. In 1917 he was awarded the Adams Prize of the University of Cambridge for a superb essay entitled *Problems of Cosmogony and Stellar Dynamics*, which was published in book form in 1919. This contained his fundamental researches on the forms of equilibrium of rotating, gravitating masses, both compressible and incompressible, and their application to the astronomical universe, together with an exposition of the relevant works of other writers. This will be described in detail in a later chapter. It suffices here to say that this volume has become a classic of astronomy. It takes a fundamental problem, which had attracted the attention of mathematicians of the first rank from the days of Maclaurin onwards, solves it in masterly style, and then proceeds, against this mathematical background, to survey the principal astronomical formations disclosed by observation and to seek out their probable origins. Its main conclusion was that the type of our solar system was likely to be very

* I find later that in 1923 he wrote one paper in the *Proceedings of the Royal Society* on α- and β-rays.

uncommon in space, that in fact our own solar system might be unique, since its origin could only be attributed to an exceptional set of circumstances—the improbable event of the near approach of twin stars. Though many of the actual conclusions of this book have received revision in the light of later discoveries—the book was writtten before the phenomenon of the expansion of the universe was known—its fundamental mathematical results form a hard core which maintains its value to this day. It is written, even in its most formidable portions, in a splendid style: great matters are treated of in great language. It must have had a decisive influence in attracting young workers to theoretical astronomy. We see a gifted mathematician marshalling his forces, surveying the field, uncovering the mistakes of his predecessors, devising new methods to make new attacks on intractable problems, and then turning aside from mathematical severities to make a grand synthesis of theory and observation.

The following letters of Jeans to G. E. Hale, Director of the Mount Wilson Observatory, may be inserted at this point:*

8, *Ormonde Gate, Chelsea, S.W.*3
Oct. 11, 1917

Dear Professor Hale,

I am engaged, as far as times permit, in preparing a book on Cosmogony for the Press, on selective [*sic*] photographs. I am not surprised to find that all the sixteen which I should like to have permission to reproduce come, without exception, from Mount Wilson.

I feel a little embarrassment in asking for permission to illustrate my book entirely from Mount Wilson photographs, but venture to do so, as yours are pre-eminently the best.

Of the sixteen which I should like to have permission to reproduce, six are by Ritchey and ten occur in the recent very fine collection by Pease in the *Astrophys. Journal.*

[Then follow details of the photographs in question.]

* I reproduce them by the courtesy of Dr W. S. Adams.

I am especially interested in Pease's edgeways spirals as they agree very well with the figures I have calculated for rotating gases. [Bakerian Lecture, Royal Society, 1917; this had not been printed yet.]

If you could give me permission to reproduce these, of course with full acknowledgement, I should be most grateful. I will write separately to Ritchey and Pease also.

My book is being produced in their best style by the Cambridge University Press (Royal 8vo format, similar to Darwin's *Collected Works*, etc.) and I am sure they can be relied on to do justice to your fine photographs. They ask for actual photographs if these can be supplied. (I have prints of Ritchey's but not of course of Pease's.)

Yours very truly,

J. H. JEANS

8, *Ormonde Gate, Chelsea*, S.W.3
Jan. 1st, 1918

Dear Professor Hale,

I am greatly obliged for your kind letter of Nov. 3 giving me permission to use the Mount Wilson photographs. I have been waiting in the hope of receiving prints from Pease, but am now beginning to fear they must have been sunk or gone astray, so I will write him again.

I am so very glad to hear that the 100-inch telescope is so satisfactory. Besides what you wrote me I heard a glowing account of its success from Dyson, who had just been receiving letters from Mount Wilson. I was greatly interested in the Andromeda nebula results which Dr Adams was good enough to send me.

Again thanking you, and with the Greetings of the Season,

Believe me,

Yours sincerely,

J. H. JEANS

In 1917, too, Jeans was invited to deliver the Bakerian Lecture before the Royal Society, and he chose as his title,

'The configurations of rotating compressible masses'. This was published as a memoir in the *Philosophical Transactions* of the Royal Society in 1919; the substance of it occurs also in his Adams Prize Essay.

It is odd that Jeans did not join the Royal Astronomical Society till 1909, presumably on his return from Princeton. It is still more remarkable that he did not publish a paper in the *Monthly Notices* of the Royal Astronomical Society till 1913. He had previously used the *Philosophical Magazine* or the *Proceedings* of the Royal Society for his papers on radiation theory and equipartition of energy, and the *Proceedings* or *Philosophical Transactions* of the Royal Society for his cosmogonic papers. But from 1916 to 1928 he was a frequent contributor to the *Monthly Notices* of the Royal Astronomical Society.

It was about this time (1917 and the immediately succeeding years) that Eddington began to publish his now well-known researches on the equilibrium of the stars, in particular his work on the radiation equilibrium of gaseous stars. Eddington got on to this subject in an indirect way: he wanted to investigate the theory that Cepheid variable stars were stars undergoing radial pulsations, but to calculate these he needed first an equilibrium theory of the stars. He soon found difficulties in applying a simple hydrostatic theory of equilibrium under gravity alone, and was led to his brilliant discovery that the pressure of radiation was an important factor in equilibrium—it had previously been regarded by physicists as an interesting but practically unimportant phenomenon. Eddington showed also that the principal mechanism of the transfer of energy would be by the emission and absorption of radiant energy by contiguous portions of the star, and, assuming the gas laws to hold good in the interior of a star, he was led to a formula expressing the absolute luminosity of a star as a function of its mass and the molecular weight of the gas composing it. He also calculated the internal temperature of the star. In these calculations he made great use of Emden's results on the polytrophic equilibrium of spheres of gas.

Jeans was much interested in these results and made a suggestion of value which Eddington at once adopted: Jeans pointed out that, at the temperature calculated by Eddington, the atoms of the gases composing the star would be highly ionized, that is, stripped of many of their outer electrons, and that in consequence, even for elements of large atomic number like iron, the *mean* molecular weight (that is, the average mass per discrete particle, counting the loosened electrons separately) would be much smaller than Eddington had supposed and of the order 2. This affected in turn the calculation of the relative importance of radiation pressure (the value of $(1-\beta)$ in a certain famous quartic equation of Eddington's involving the mass of the star and the mean molecular weight as coefficients) and, in turn again, the calculation of luminosity.

Thus Jeans had no prejudice against Eddington's theory. Indeed Jeans himself needed just such an equilibrium theory as a basis for his discussion of *rotating* compressible masses. But, reflecting on Eddington's main result, that the luminosity of a star of given mean molecular weight could be calculated in terms of its mass only, without discussion of the physical origin of the energy liberated in the interior, Jeans began to be puzzled. The result was a series of papers in the *Monthly Notices* in which he outspokenly criticized Eddington's theory.

In setting up his equation of transfer of radiation, Eddington had had occasion to say: 'Let $4\pi\epsilon$ be the energy liberated per unit mass in the star's interior.' But this symbol ϵ made no appearance in Eddington's final luminosity formula. Jeans rightly objected to Eddington's conclusion that the luminosity was calculable in terms of the mass; he contended that this luminosity was precisely $4\pi\epsilon M$, where M was the mass of the star, and that a star of given mass M could have any luminosity whatever, depending on the value of ϵ. Eddington's result was therefore correct, but trite, and he had established no result of any value.

There was an answer to this contention of Jeans's. What Eddington had actually calculated was not ϵ or $4\pi\epsilon M$, but L, the rate of radiation from the star's surface, and the star would only remain in a quasi-steady state if its interior were endowed with sources of energy, of amount $4\pi\epsilon$ in the mean, just totalling to L, so that $L = 4\pi\epsilon M$. Eddington's luminosity formula was in fact simply the boundary condition that the configuration should remain as a globe of perfect gas throughout. This is further borne out by the circumstance that Eddington used a comparison of his formula with observation to yield a numerical value for what he called the *internal opacity* of the star (the coefficient of absorption of radiation), but what was in fact the numerical value of the surface opacity.*

Eddington, however, did not make this reply. He appeared to take umbrage at having his results criticized at all, and wrote:

> Instead of proceeding in this way [i.e. instead of evaluating L in terms of mass, opacity and molecular weight] Mr Jeans reverses the steps of my argument, and arrives at the result that the total emission is $4\pi\epsilon M$, which, he remarks, is right but obvious, so that the discussion does not advance our knowledge of stellar conditions in the least. He is not likely to advance our knowledge by undoing my work. He merely verifies my algebra. The unknown quantity ϵ was introduced near the beginning of my paper, and eliminated as soon as convenient; the main thing accomplished in my paper was the determination of its value in terms of k, ρ, M, etc. Mr Jeans throws this aside, and resurrects the original symbol.

This was all very unfair to Jeans, who was an honest critic, honestly expressing his difficulties. But Eddington

* See my Presidential Address to the Royal Astronomical Society, *Monthly Notices* (1945), 105, 146.

loved to make debating-point replies, as the present writer found when he also criticized Eddington a decade later. A little later, Eddington made another debating point, which confused the issue further. Eddington in the same paper wrote:

Under very general conditions I find the rate at which a giant star of given mass and opacity radiates energy. The star can only be in a steady state if energy from some source is supplied at this rate. If it is difficult to imagine a source supplying energy at just the right rate, that is an argument against the steady state, not against my theory. Jeans's hypothetical star of intense radioactivity will not necessarily radiate faster than an ordinary star; it will simply use the surplus energy in expanding.

This again was an illegitimate argument. It is clear that, in general, an arbitrary mass M, endowed with sources of energy totalling to an arbitrary amount L, will find a steady state for itself if L is not too large. Whether this is a purely gaseous configuration or not is just the point that ought to have been the subject of investigation. What Eddington had really done was, not to calculate the luminosity of a star of arbitrary mass M, but to find the precise luminosity needed to ensure that it was in equilibrium in the form of a perfect gas throughout. Since it was inconceivable that the sources would supply energy at the precise rate required, Eddington's argument *prima facie* was a demonstration that the stars could not be built on the model considered.

Jeans realized that Eddington's result was closely connected with the surprising properties of the Emden polytrope '$n=3$'. This configuration has no determinate radius, and a slight departure from the precise contribution of radiation pressure necessary to hold it in equilibrium has catastrophic effects on its form. If radiation pressure is slightly in excess, it has to develop a singularity (infinity) of density at its centre in order to maintain equilibrium; if radiation pressure is slightly in deficiency, it can only exist in the polytropic form '$n=3$' with an internal supporting surface. The latter would yield in nature a 'collapsed' configuration of some new type, since nature could provide no internal supporting surface. Thus the real conclusion to be drawn from Eddington's investigations was that the stars in nature could not be built of perfect gas throughout.

Unfortunately Jeans misled himself as to the origin of Eddington's own conclusions, and thought he had detected an error in Eddington's derivation of the equation of radiative transfer. It is true that there were obscurities in Eddington's original proof of this equation, when sources of strength $4\pi\epsilon$ were present; but there was no actual error of approximation, as Jeans accused Eddington of making (*Monthly Notices* (1917), **78**, 37). However, good came out of Jeans's misunderstandings of Eddington's approximations, for the paper which Jeans wrote to clear the matter up (*Monthly Notices* (1917), **78**, 28), contains some valuable analysis on the theory of radiative equilibrium in the boundary layers of a star, and inspired a good deal of later research by others. It also found an approximate value of the coefficient of darkening of the disk of a star towards the limb.

No words are needed to praise Eddington's achievement in calculating the state of equilibrium of a given mass of gas, and in calculating the rate of radiation from its surface. What was wrong was Eddington's failure to realize exactly his *achievements*: he had found a condition for a star to be gaseous throughout; by comparison with the star, Capella, he had evaluated the opacity in the boundary layers; and he had made it appear unlikely that the stars in nature were gaseous throughout. His *claims* were the contrary: he claimed to have calculated the luminosity of the existing stars; he claimed to show that they were gaseous throughout: and he claimed to have evaluated the *internal* opacity of the stars.

Jeans deserves great credit for being the first critic to be sceptical about these claims of Eddington's theory, in spite of the attractive plausibility with which the theory was expounded. I think that even today there is much misconception amongst astronomers about the status of Eddington's theory. The tenacity with which Eddington hung on to his first ideas and declined to modify them as research and understanding progressed, coupled with the extraordinarily attractive style of his work, *The Internal Constitution of the Stars*, has been responsible for the slowness with which astronomers have been able to winnow the chaff from the grain in his work. The present writer in 1929 became sceptical of the validity of Eddington's own account of his conclusions, and when illumination came, it came with the shock of a revelation, of a sudden conversion (or anti-conversion!). Jeans was indeed wrong about the origin of Eddington's mistranslation of his mathematical work; but he had the courage to express his *malaise* as to the legitimacy of the results. It is much to be regretted that these two Titans, Eddington and Jeans, should not have co-operated in their assaults on the grand subject of stellar structure, instead of being opposed to one another, during the most constructive periods of their careers. The blame has to be divided between them. Jeans mistakenly attacked Eddington's mathematics instead of accepting his mathematics and then providing the correct interpretation; Eddington resented what he considered to be aspersions on his competency as a mathematician, and never understood the difficulties of a philosophical kind that surrounded his own interpretations of his results.

Astronomers on the whole have favoured Eddington's side of the controversy—mistakenly, in my opinion. This is due, in addition to the reasons mentioned above, to the fact that Eddington had more of a feeling for the physics of a situation than Jeans had, whilst Jeans had more of a feeling for the mathematics of a situation than Eddington had; the result was that Eddington's stars had a physical plausibility that Jeans's lacked, and the astronomer who did not wish to go into the rights and wrongs of the mathematical situation could see the physical likelihood of Eddington's being correct. Stars *do* behave as though, for the most part, their luminosities are a function of their mass; Eddington discovered this empirically, though guided by a mistaken view of his own mathematical results. Again, when the central temperature and density of a star have been estimated, it seems not unlikely that comparison of luminosity with mass will yield a numerical value of the internal opacity; the apparent verbal argument is that the opacity in the interior acts as an obstacle to the flow of radiation, and thus naturally appears in the form of a quotient of mass (or some function of the mass) by luminosity. Actually the observed luminosity of a star, being the value which the flux of energy takes at the boundary, is the rate at which the gaseous mass, in its present condition of temperature and density, is cooling to space, and it thus determines the boundary value of the opacity.

Jeans, in contending that M and L are *independent* variables as far as equilibrium considerations are concerned, was correct; but he should have gone on to display the various stellar quantities—radius, central temperature, density and pressure—as functions of these two formally independent variables L and M. This he never did—at least explicitly.

When in 1929–30 I independently attacked the problem of stellar structure, and succeeded in representing the various stellar quantities as functions of the two independent variables L and M, I lost all faith in the physical validity of Eddington's conclusions about the stars, and in a debate at the Royal Astronomical Society I said openly that in my opinion Jeans had been right in his contention regarding the physical basis of Eddington's theory. It happened that de Sitter, of Leiden, was present at that debate, and he came to Oxford next day (I think), as my guest. Whilst staying with me he remarked that *scientific controversies were not settled by taking sides.* In his courteous way he intended thus to rebuke me for my remarks at the debate, and I took his remarks in this sense. He made me feel, somehow, that I had committed some impropriety, though actually my remarks had been perfectly honest and the consequence of my own mathematical investigations. However, I decided not to lend myself to any similar misconstruction on future occasions, or to appear to bid for Jeans's support—Eddington had as usual refused to listen to the nature of the criticism I was making—and so at the resumed debate of January 1931, on stellar structure I rather unwisely went out of my way to attack Jeans's own theory of the internal constitution of the stars, which he had developed in the intervening decade, on the ground that his hypothesis of the presence in stars of atoms of very high atomic number was outside physics. This was, however, only after I had stated my agreement with Jeans that in the equilibrium theory of the stars L and M were independent variables. Jeans in his reply welcomed my agreement with him on this point but, in his own words, criticized my work as bluntly as I had criticized his. He did not see that it went further than his own work of 1917–19. When the time came for me to reply to the debate, I rebutted Jeans's criticisms, and said that I had been doing in my own papers what he had not done in 1917–19, that I had performed the task he ought to have performed in 1917–19. I fear I was rude to Jeans in this debate; whether he was rude in reply can be left to the judgement of others: the whole debate is recorded in the February number of the *Observatory* for 1931. This was the only occasion on which I was ever in open contention with Jeans; bystanders of the debate spoke afterwards of our having 'wiped the floor' with one another. But when I met Jeans a little later at a gathering of one of the scientific societies, Jeans was courtesy itself. Indeed, I always experienced the greatest kindness whenever I had dealings with him.

The outspoken astronomical debates between Jeans and Eddington of 1917–19 had one interesting consequence. Hardy, Jeans's competitor in the tripos and the Smith's Prize, became a Fellow of the Royal Astronomical Society, with some others, in order to have the privilege of attending these debates, and hearing Eddington and Jeans castigate one another in public. And sure enough, Hardy attended the debate of January 1931, in due course, when the conflict had become triangular, and, moreover, contributed to it. His contribution was characteristic: he was presenting some analytical results obtained by R. H. Fowler (later Sir Ralph) in connexion with the fundamental differential equation of stellar structure, and he remarked that, amidst all the contending theories of Eddington, Jeans and Milne, of one thing he was certain, namely that when these theories had all become dead and forgotten, the pure mathematics dealt with by R. H. Fowler would survive.

The contentious papers of Jeans and Eddington and the wide interest and genius of their constructive ones appreciably increased the circulation of the *Monthly Notices* of the Royal Astronomical Society in which they were published.

It should be added that during their lifetimes neither Jeans nor Eddington ever budged from the positions they had taken up *vis-à-vis* their respective papers on stellar structure. But this opposition did not extend to papers on other, even if closely allied, subjects. They tacitly agreed, evidently, not to refer to the thorny subject on which they disagreed.

CHAPTER IV

SECRETARY OF THE ROYAL SOCIETY
1919–29

In 1919, Jeans was awarded one of the Royal Medals of the Royal Society, and in the same year he was elected to the office of Honorary Secretary of the Society. Hale's letter of congratulation to Jeans contains several points of interest:

Pasadena, California
May 25, 1920

Mr J. H. Jeans,
Cleveland Lodge, Dorking, England

My dear Mr Jeans,

I meant long since to send you my heartiest congratulations on your election as Secretary of the Royal Society, or rather to offer my congratulations to the Council, as I am sure that the event is of no small significance in its bearing on the future development of the Society.* This is unquestionably a very critical period in the progress of science and the policy adopted by such authoritative bodies as the Royal Society may turn the scale in the right direction. In this country, and probably England, the conditions are very complex. On the one hand, the increased cost of living and the high salaries offered by the industries are drawing good research men away from the faculties of educational institutions. On the other hand, there is such a marked advance in the public appreciation of science and research and such an obvious necessity of developing more investigators that the opportunity to interest governments and individual donors is greater than ever before. This is manifested in part by the strong expressions of the value of pure science made by industrial leaders. The pamphlet I am sending you under separate cover (*Scientific Discovery and the Wireless Telephone*

* Hale had been elected a Foreign Member of the Royal Society in 1909.

was prepared by the American Telephone and Telegraph Company to accompany their exhibit of the wireless telephone and its scientific development, first shown at the building of the National Research Council in Washington and now at the American Museum of Natural History in New York. You will see what emphasis they lay upon the importance of research in physics without reference to practical return. If we can convince everyone of this, I am sure we can obtain large new funds for pure science.

Just at present we are attempting to secure a fund of a million dollars for research in physics at the California Institute of Technology (in Pasadena) and from the progress made during the last few days I hope we may complete it before July first. I have had no difficulty in convincing the trustees of this small institution (about 300 students) that research and graduate instruction should have quite as prominent a place in their programme as the ordinary undergraduate work of the average school of technology, and the prospects for the future are very encouraging especially as we mean to limit the attendance in order to raise the quality of the work. It is of course true that the trustees must be given clear ideas of the meaning and importance of research, but there should be very little difficulty in educating any average board, provided one who is familiar with research and its recent developments can have occasional opportunity to present the arguments.

I hope you are going ahead with the organization of National Committees of the various International Unions, and that these may perhaps be ultimately united to form a National Research Council, as I understand from Lacroix (if I have deciphered his letter correctly!) that the French intend to do. American Sections of the Astronomical, Geophysical and Chemical Unions have been formed, and that of the Mathematical Union is now in process of organization. I particularly hope that you are forming a Section of the Mathematical Union, and will be well represented at Strasbourg. Our mathematicians felt that the French went ahead rather rapidly, and are concerned over the fact that the date selected for the Strasbourg Congress coincides with that of the opening of our universities and falls in the same month with the regular triennial mathematical colloquium, to be held this year at the University of Chicago. But in the general interest they have smothered their chagrin, and various national societies are joining in the organization of a strong Section, which plans to send delegates to Strasbourg, where they expect to present a scheme for an abstract journal of mathematics, which seems to be greatly needed. They will be much encouraged if England is well represented there, for in spite of political squabbles and the infernal machines of Sinn Feiners, Egyptian nationalists, Indian Fakirs, and the unspeakable Hearst, England and the United States must work together and do everything possible to put down the many elements that feed on discord.

We are now at work on the definitive observing programme of the 100-inch telescope, and it is hardly necessary to say that your book, which I admire so much, is our chief guide in preparing the attack on the spiral nebulae and on many other questions. I hope you will send me suggestions from time to time, as our programme will be kept elastic enough to follow them.

Give my kindest regards to Mrs Jeans, and believe me,

Yours sincerely,
G. E. HALE

All secretaries of societies, if they are efficient, have some influence upon policy and Jeans was responsible for a very notable improvement in the status of the Society's *Proceedings.* He made them into what may be justly called the premier English journal for the publication of researches in the physical sciences. Previously the *Philosophical Magazine* had been accustomed to receive the most important original papers in physics (e.g. by Rutherford and his pupils, Bohr, Darwin and others, and most of Jeans's own papers in theoretical physics), while the more monumental essays were published in the Society's *Philosophical Transactions* ('buried', as Rutherford used to say, 'in the mausoleum'). Rutherford was President of the Royal Society from 1925 to 1930 and with his backing Jeans set out to capture the best papers in British physical science for *Proceedings* A. What he set out to do, he achieved.

Though the work of appointing referees for papers was nominally the work of a committee in each subject, nearly all the papers published in *Proceedings* A must have passed through Jeans's hands and he was frequently helpful to young authors.

Further honours came to Jeans at this time. He was awarded a Hopkins Prize of the Cambridge Philosophical Society for the period 1921–4 'for his work on the theory of gases and on radiation, and on the evolution of stellar systems'; and he was made a Research Associate of the Mount Wilson Observatory in 1923—a rare honour, which he valued highly. In the letter of invitation George Ellery Hale wrote:

Geneva, Sept. 9, 1924

Dear Dr Jeans,

I was greatly disappointed to miss you in England, especially as I cannot hope to return there this year and probably not during this trip to Europe. The whole trouble lies in the weak condition of my head, which was greatly aggravated in London by the curious stimulus and excitement regarding scientific matters I invariably feel there. I had not failed to appreciate your very kind letter received in Pasadena, or the cordiality of your invitation to visit you in Dorking, which I remember so very pleasantly. But I had reserved my reply until I could see more clearly the possibilities of travel, which depend so directly upon the variable, and usually unpredictable, state of my head. I intended, for the sake of much needed rest, to travel for some time in Devon after landing, but the continual rain drove us to London. As soon as I arrived there and felt the old stimulus my head was so quickly and constantly congested that I had to leave town again. On my return I missed you twice at the Royal Society, and my telegram was a last effort, the best I could make under the circumstances. As a matter of fact, I have been told by my physician to avoid all scientific discussions which tax me severely, though I can still do a little quiet writing if it involves no hard thinking. Pardon me for these details, but I owe you an explanation for failing to get into touch with you earlier.

Through an error in forwarding, your welcome letter in reply to my telegram has only just reached me here. I fear you may have already left England, as I was told that you were going to the United States, so I am requesting that this letter be forwarded, and trust that it will reach you soon.

I wanted to discuss many questions with you, but one in particular requires early consideration. Kapteyn, as you know, was a Research Associate of the Mount Wilson Observatory. This means that he received an honorarium of two thousand dollars per year but was under no obligation whatever other than to publish in our *Contributions* occasional papers dealing with problems arising at Mount Wilson. He spent several summers with us, but he was free to come or not as he chose. In short, the chief principle involved in the position is that of establishing friendly co-operation, without hampering restrictions.

I am not sure whether the position held by Kapteyn will be maintained, as the Carnegie Institution has reached, if not passed, the full limit of its income, and the appropriation to the Observatory for the coming year has not yet been fixed. If, however, I am free to make the appointment, I should like very much to offer it to you. If you honored us by accepting, I should be greatly pleased, especially as I feel sure that co-operation would be highly advantageous to us and possibly of some use to you. We should greatly value your occasional suggestions regarding new problems or special lines of observation, and we should be glad to place before you certain questions that puzzle us, though you would never be under any obligation to reply to them if the answers were not evident to you. The purpose of the honorarium is not to pay the Research Associate for work done for us, but to contribute in a modest way towards the pursuit of his own researches in the general field in which we are interested.

I need hardly say that whether you do or do not see your way clear to accept such an offer, assuming it can be made, you will be more than welcome to visit the Observatory at any time and to stay there as long as you choose. I am sure you would find many points of importance to discuss with members of our staff and with Millikan, Epstein, Noyes, Tolman and other investigators now at the California Institute of Technology in Pasadena. If you go to Pasadena on this trip I shall unfortunately be compelled to miss you, as I am condemned to stay away for at least a year. But it might be more convenient to you now than later, and I hope you will follow your own inclinations, remembering that many of our men will always be able to be of more direct service to you than I, especially as my head shows no prospect of any marked improvement in the near future.

Hoping for a favorable reply, and regretting very much that I missed you, I am, with kind regards to Mrs Jeans,

Yours very sincerely,
G. E. HALE*

Jeans delivered the Guthrie Lecture of the Physical Society of London in 1923. In 1922 he had been awarded the Gold Medal of the Royal Astronomical Society, for his contributions to theories of cosmogony. The medal was presented, it is pleasant to record, by his former antagonist Eddington, who delivered on the occasion one of his characteristically eloquent addresses (*Monthly Notices*, (1922), **82**, 279). Eddington first dealt with Jeans's work on rotating masses of fluid (see Chapter x below), referring especially to the conclusion that the birth of a solar system is astronomically an exceedingly improbable event:

By one of the ironies of science the study of the break-up of a mass by rotation has taken us on a tour round the universe, finding an application seemingly to every cosmogonical problem except the one which it was originally called in to explain, viz., the development of the solar system. As to the solar system, I will quote Dr Jeans verbally: 'It is probable, but by no means certain, that we must abandon the nebular hypothesis of Laplace.'

Eddington went on to describe the tidal theory, favoured and largely developed by Jeans in his Adams Prize Essay,

* Transcribed by courtesy of Dr Walter S. Adams, late Director of the Mount Wilson Observatory.

according to which the solar system of planets was brought into being by a close encounter between our sun and a passing star, which raised tidal waves in our sun and subsequently disrupted it, the sun shedding off an arm which condensed to form the planets. The probability of such a close encounter was, on Jeans's calculations, exceedingly small, and hence it was exceedingly unlikely that any appreciable fraction of the stars in general are accompanied by planetary systems. Eddington went on:

I suppose that nothing in astronomy has appealed more to human imagination that the conception of each of the myriad points of light in the sky as a sun giving warmth and light to an attendant circle of planets....It has seemed a presumption, bordering almost on impiety, to deny to them inhabitants of the same order of creation as ourselves. But we forget the prodigality of Nature. How many acorns are scattered for one that grows into an oak? And need she be more economical of her stars than of her acorns?...If indeed she has no grander aim than to provide a home for her pampered child Man, it would be just like her methods to scatter a million stars whereof but two or three might happily achieve the purpose.

Let me repeat that Dr Jeans does not claim to have *established* that our solar system is a freak system, unusual and possibly unique in the universe; but he has shown how shaky is the opposite view which expects a system of planets to be a normal appendage of a star.

Eddington could not deprive himself of the fun of slightly teasing Jeans on the subject of their disagreements on the subject of stellar structure, which have been described in the last chapter:

Connected with this part of the subject are Dr Jeans's contributions on the radiative equilibrium of stars. I will not give details, because this has been the subject of vigorous controversy between us; and on this occasion our long-suffering Medallist is forbidden to defend himself. The *Monthly Notices* record how we have hurled at each other mathematical formulae—the most undodgeable of missiles when they are right—and the onlooker will perhaps conclude that *someone* was badly annihilated. But it is possible that Jeans and I may still have a difference of opinion—as to precisely whose corpse lies stricken on the field.

Eddington concluded by a powerful account of Jeans's researches in stellar dynamics, not yet referred to in this volume. They were published in papers in the *Monthly Notices* from 1913 to 1916, and constitute an application and development of methods devised for gas dynamics of stellar systems. They are described below (Chapter x) and form some of the most beautiful and permanent of Jeans's mathematical researches.

Jeans was elected President of the Royal Astronomical Society for the two years 1925–7, and had occasion to deliver three addresses in presenting the Gold Medal of the Society to Sir Frank Dyson (1925), to Albert Einstein (1926), and to Frank Schlesinger (1927). All three addresses illustrate the felicity of Jeans's style on a formal occasion and deserve quotation.

The award to Sir Frank Dyson, Astronomer Royal, was 'for his contributions to Astronomy in general, and, in particular, for his work on the Proper Motions of the Stars'. Jeans began as follows:

It has not been the custom of the Society to award its Gold Medal to Astronomers Royal, indeed, only one holder of the Greenwich Office has been so honoured before, namely Airy, who received the medal first in 1833 and again in 1846, not to mention receiving one of the testimonials in 1848 which fell in showers, alike on the just and the unjust, on the discoverers of Neptune, on those who might have discovered Neptune if they had been more alert, and on many at least who could never by any chance have discovered Neptune. The reason why Astronomers Royal seldom achieve medals is not difficult to discover; it was expressed with the utmost clearness from this Chair on the last occasion on which the medal was presented to a Greenwich Astronomer Royal. In presenting the medal to Airy in 1846, Captain Smyth said: 'Our medal was primarily instituted as a mark of approbation on individual exertion, on labours of love; and not to note our sense of the official merits of public men, or of the rectitude and ability with which they may acquit themselves in their respective offices.'

An Astronomer Royal must of necessity be a busy man; it is naturally somewhat rare to find one who, after the accomplishment of his heavy routine duties, has either the energy or the inclination for astronomical labours of love. It is because the present occupant of the office is such a one that it is my pleasant duty to present the medal to him today.

Jeans went on to describe Dyson's labours jointly with William Grassett Thackeray, under the direction of Sir W. H. M. Christie (then Astronomer Royal) in determining and publishing in 1905 the proper motion of 4239 stars by comparison of Stephen Groombridge's observations (1806–16) with Greenwich observations and others, a volume which allowed confirmation to be made of Kapteyn's discovery of the two star-streams. Dyson subsequently showed that Schwarzschild's ellipsoidal law of distribution fitted the observed facts better than Kapteyn's two-stream hypothesis. Jeans referred to this in one of his powerful metaphors:

At first sight the two hypotheses of Kapteyn and Schwarzschild seem to be in irreconcilable opposition, but I think it is possible that the distinction between the hypotheses is after all one of degree. Armies of stars may be intermingled in the way imagined by Kapteyn, but instead of fighting their way through one another gravitational forces may provide attractions which result first in spasmodic fraternizations between members of the opposing armies, and finally in the two armies, now welded into one, marching through space as a single army, in which case it would exhibit precisely those peculiarities of motion which were suggested by Schwarzschild.

Earlier in the address, Jeans had remarked prophetically:

The question of whether star-streaming is merely a local phenomenon or extends through the whole of our galactic system is of the utmost importance; it cannot yet be fully answered, but the answer when it comes will probably bring with it the solution of the puzzle of the structure and origin of our system of stars.

The theoretical explanation of star-streaming was ultimately given by Bertil Lindblad of Stockholm, as consisting of the residual motions left in the galactic system when the main contribution of galactic rotation had been removed.

Jeans complimented Dyson on the way in which he had sunk his personality in his observatory....It is not generally possible for all the observers to have the wide vision of the Director, but Dyson has, I know, endeavoured, and with success, to make each observer keep the object of his observations always in view. They have not been treated as mere hewers of stones, told to cut a stone to such and such a shape, but rather as co-architects, with whom the director has freely and frankly discussed his plans for the bit of the structure immediately in hand, and in consultation with whom he has also tried to envisage the whole edifice as it will finally appear.

Jeans also pointed out that in becoming a devotee of proper motions, Dyson had robbed the world of a first-class spectroscopist, as evidenced by his eclipse observations of 1900, 1901 and 1905. Lastly, Jeans spoke of Dyson's part in organizing the eclipse expeditions which culminated in the observation, in 1919, of the deflexion of light by the sun's gravitational field as predicted by Einstein's general theory of relativity. This, said Jeans, made on him the greatest impression of all.

The second address was on the occasion of the presentation of the medal to Einstein 'for his researches on Relativity and on the Theory of Gravitation'. Here again Jeans began with a charming simile:

Some of us may remember the story of the children who played truant in order to explore the regions where the rainbow ends. After travelling all day, up hill and down dale, they had to admit failure of the most thoroughgoing kind—the rainbow was, to all appearances, no nearer than when they started.... They must have felt they were the victims of extremely bad luck, for they had clearly seen the rainbow in front of the nearest hill when they started out; could there be some sort of conspiracy on the part of rainbows, hills, and indeed the whole scheme of nature, to prevent their getting close to that rainbow?

In the year 1905 the world of physicists was engaged in a pastime which was, in many respects, very similar to that of chasing rainbows. They believed light to travel through an ether with a speed of 300,000 km. a second. If the solar system were travelling through this ether at a speed of, say, 1000 km. a second, it would partially overtake light travelling in the same direction, so that this light ought to appear to travel at only 299,000 km. per second. On the other hand, light travelling in the opposite direction ought to appear to travel with a velocity of 301,000 km. a second. This suggested an obvious means of discovering both the speed and direction of the earth's voyage through the ether....Experiments were designed to utilize this principle, and were tried time after time, not in one form only but in many. And time after time, experiment gave the answer that the velocity of the earth through the ether was zero....No doubt it was conceivable that on the occasion of the first experiment the earth really might have happened to be at rest in the ether; but it was quite inconceivable that this should be the case every time; indeed the earth's orbital motion alone required a variation of some 30 km. a second, and all the experiments were capable of detecting far smaller variations than this. At first glance it looked as though the earth must be carrying its own private ether about with it, and I suppose this view would have prevailed had it not been for the astronomers, who were ready with an aberration constant which at once, and I think irrevocably, dismissed the possibility of an ether being dragged about with us....

I forget the end of the rainbow quest, but am prepared to provide an entirely unauthoritative ending. After the children had got completely tired in their bodies, and still more completely bewildered in their wits, they rested for a long time, until they encountered a magician. He was not in the least the conventional magician, ponderous of speech and with a long white beard; indeed he was a young man of twenty-six, extraordinarily simple and unassuming in all that he said. What he said in brief was this: 'I can tell you what is the matter. You have started to chase the rainbow on the supposition that it is a material arch; in actual fact it is all in your own eyes. Gretchen sees one rainbow, and Hans an entirely different one. But if Hans walks up to where Gretchen is standing, he simply changes his rainbow for hers; you don't get any nearer to a rainbow by walking this distance, because there isn't really anything for you to get any nearer to....As the children were tired, and the young magician had expressed himself in rather unfamiliar ways, they didn't at first quite understand what he meant. But then another magician whose name was Minkowski came along, and he made it all seem much simpler; he said it was quite true that each child carries its own rainbow about with it, but that behind the subjective vision of the rainbow was an objective reality consisting of a shower of raindrops. These raindrops were the same for everybody, but out of the whole lot each person's eye selected, or rather the sunshine selected for each person's eye, a small group of drops which appeared to him to form a bright arch. If all space were filled with children standing in different spots, then the aggregate of all the raindrops seen in all the children's eyes would constitute the reality behind the phenomena, a shower of rain. When the second magician put things in this way, the children began to understand; they saw that the first magician, whose name was Einstein, had been right.

I doubt if the Society ever had its Medallist introduced to it in so disrespectful a way before, but my little parable may remind you of the way in which our present Medallist made his entry into the scientific world, and also of the way in which the scientific world made their entry into the changed universe in which science moves today. Time and space as separate entities, the time and space we wrote about and thought about previous to 1905, have gone, or, as Minkowski puts it, have become shadows, while only the product of the two remains as the framework in which all material phenomena take place.

Jeans's whole address is couched in equally striking and simple language. One need not agree with the details of Jeans's account of Einstein's work, or even assent to the correctness of Minkowski's views, to savour and appreciate the wine of Jeans's language, and its bouquet. And it must be remembered that this was before Jeans had become what one may call a 'professional' writer of popular science. In the concluding part of the address, as so often in Jeans's utterances, there is a highly prophetic sentence (it must be remembered that this was February 1926): '...there is now, for the first time since Newton, room in the universe for something besides predestined forces.'

At the same time I think that we must notice here what I have more than once noticed about Jeans's writings, that when dealing with work to which he had not himself contributed, he became less critical, indeed almost complacent. It was not merely that Jeans so fully appreciated the work of others that he could not see their limitations: we have seen that in the case of Eddington's work, and in the case of Planck's work, Jeans approached both with a usefully sceptical outlook, and consequently succeeded in enriching their subjects. But this sceptical attitude forsook him when dealing with relativity, or the more recent advance in atomic physics. Minkowski, in fusing together time and space, lost sight of the character of time essential for the observer, namely its one-way-ness. Einstein, in using a curved space-time as the scene of gravitation, reintroduced in effect an ether or absolute background which he had done so much to expel from physics. But Jeans rarely hints at these difficulties.

Unfortunately Einstein could not be present in person on this occasion, though he sent a short and characteristically modest letter of thanks, which Jeans read at the conclusion of the presentation.

The third of Jeans's addresses in this series was on the award to Frank Schlesinger for his work on stellar parallax and astronomical photography. Jeans's extraordinary versatility is seen at its best in this address, for its subject was a technical astronomical achievement in a field in which Jeans could not have been supposed to have special knowledge, or be specially interested. Yet Jeans makes the whole subject read like a fairy story. His exordium was as follows:

At the dawn of civilization, when man awoke from his long intellectual slumber, nine muses were appointed to preside over his various activities. Only one muse was allotted to science, and that one was allotted to astronomy. Perhaps the high gods of Olympus who arrange these things had heard of no science beyond astronomy, or perhaps they thought it the only one worth encouragement; we do not know. But it is said that since then the claims of other sciences have been admitted, and a vast crowd of junior muses have been appointed to look after them. When they all meet in conclave, many of the latter report quite extraordinary rates of progress in their particular sciences, and it has sometimes been thought that astronomy, which started first in the race, has at times shown some tendency to lag behind.

To this Urania has always had a ready answer. She can point first to the immensity of her task; sciences such as geography, geodesy and geology, whose field of action is limited to the surface of one tiny planet, can no doubt claim to be well on towards the completion of their tasks, but the exploration of an entire universe offers a task of a different order of magnitude. She can also point to the extraordinary difficulty of this task; after the first obvious steps have been taken, the astronomer can get nothing of value except with instruments of almost incredible precision. Moreover, instead of being able to investigate phenomena when he pleases, he has to wait on Nature, and Nature moves very slowly in comparison with human life—the whole age of astronomy, as a science, bears the same relation to the ages of the stars that it studies as does the last tick of the clock in the dying century to the century itself. Finally, she has often been heard to remark, with a mixture of pride and sorrow in her voice, that her science presents problems of such enthralling interest that only too few astronomers can be found to do their fair share of the fundamental work on which all progress must ultimately be based. We can rest assured that she is well satisfied when, as on the present occasion, our Society signals out for its highest honour one who has done not only a fair share, but a lion's share, of most valuable fundamental work.

Jeans went on to trace the history of the determination of stellar parallaxes (of the absolute distances of the stars), showed how the early astronomers, believing the stars to be comparatively near, thought it would be an easy task to determine the distances of their various 'heavenly spheres' with fair accuracy; how later on, when the determination did not prove so easy as had been anticipated, the difficulty was used as a real argument against the Copernican system; how Hooke, Römer, Bradley, all attempted parallax determination and failed, though Bradley's failure was a glorious one, resulting as it did in the discovery of aberration; how Kepler had maintained that the stars were probably of the nature of suns, and how this view might have been used to show that they could have a parallax of at most two seconds of arc; how Newton actually advanced this argument; how the eighteenth-century astronomers began to realize its implications, and how in 1800 Herschel used Kepler's supposition to compute hypothetical parallaxes for bright and for faint stars; how the year 1838 saw the whole position suddenly transformed, by the almost simultaneous but independent measurements of parallaxes of three stars by Bessel, Struve and Henderson: how this advance was greeted with enthusiasm by Sir John Herschel in handing the Gold Medal to Bessel in 1841; how in spite of skilled attacks on the problem, progress with further determinations of parallaxes was very slow; how in 1901 Newcomb could compile a list of only seventy-two stars whose parallaxes had been measured, including some very doubtful ones, so that knowledge of parallaxes was growing only at the poor rate of one per year; when suddenly Frank Schlesinger (the medallist of the current year) showed the way to determine accurate parallaxes, by photographic methods, in great numbers.

Jeans concluded his account of Schlesinger's parallax work by showing that even at that date, 1927, the survey of the stars in the neighbourhood of the sun was 'complete' only to some 4 parsecs. This led him to another brilliant passage:

> Just now we compared astronomy's past 3000 years of existence to the last tick of the clock in a dying century; we may with equal justice compare the 3000 years next to come to the first tick in a new century—such is the proportion it bears to the total time our present stars may be expected to last. Looked at on this scale there may be thought to be plenty of time for a really exhaustive survey of the universe, and past progress may be accounted amazingly rapid. In one tick of the clock astronomy has come to birth, has developed the most difficult and accurate technique known to science, and has surveyed the universe, accurately and completely to 4 parsecs, and sketchily and imperfectly beyond. The next tick of the clock should see the accurate survey extended to well beyond 100 parsecs, and astronomy has a whole century of life before it. I fear there may be a fallacy involved in the adoption of this point of view: it is that we do not know how long the human race will endure. At the very longest estimate, man has only existed for a hundred ticks of our clock, civilized man for only two or three, so that our race and civilization would alike seem to be at the very beginning of their existence. And yet he would be a bold extrapolator who would assert with confidence that man is good for another three thousand million ticks on the ground that, if nothing unforeseen happens, the present stars will still be in existence at the end of that time. When we take a time-scale in terms of individual lives, we may still feel gratified at past progress, yet we cannot but feel awed by the magnitude of the task that remains.

It has seemed worth while to make these lengthy quotations from Jeans's presidential addresses to the Royal Astronomical Society since otherwise such utterances are apt to remain buried in the annual reports of the council. They exhibit Jeans's gift for arresting phraseology in dealing with technical subjects some years before this was recognized by the general public. But I have another reason for these quotations: the day of the full presidential address on the award of the Gold Medal seems to be passing away, largely as the result of an act of generosity and piety on the part of Jeans himself. To this I now come.

The *Monthly Notices* of the Royal Astronomical Society for June 1926 contains a statement to the effect that the Treasurer (Col. F. J. M. Stratton) had received the following letter from the President (Dr J. H. Jeans).

My dear Treasurer,

The Royal Astronomical Society has no lectureship or other endowment such as makes it possible in many Societies for an eminent person from abroad to be invited to give a lecture on his own work. It would give me great pleasure if Council should see fit to accept the following offer:

I offer the sum of £1000 (a thousand pounds) for the endowment of a 'Foreign Lectureship', the capital to be kept intact, and the income to be used for an annual lecture on some subject of interest to astronomers, preference being given to a lecturer normally resident outside the British Isles. I had thought that in particular the Medallist of the year, when resident abroad, might frequently be invited to be Foreign Lecturer also—it would give him an extra inducement to come in person to receive his medal. When no suitable Foreign Lecturer was available, Council could of course invite one of their own Fellows or someone in some cognate science, geophysics, or even physics of the right kind.

If Council accepts this offer and if, after fair trial of the Foreign Lectureship, they consider the money could be used better in other ways, I should of course wish this to be done. I should prefer (but without creating any legal obligation) that the income should be used for some specific purpose, and not merged in the general funds of the Society, but quite realize this might not be possible.

Will you be so good as to bring this matter before Council?

Yours sincerely,

J. H. JEANS

Lt.-Col. F. J. M. STRATTON,
Treasurer, R.A.S.

The Council gratefully accepted this offer, and appointed a small committee to consider the best way of carrying out the wishes of the donor. After consultation with him, and in accordance (I understand) with his wishes, it was decided to use the capital of the gift to found 'an annual lectureship, to be called *The George Darwin Lectureship*, the lecture to be on some subject of interest to astronomers, preference being given, in electing a lecturer, to one normally resident outside the British Isles'. (Report of Council to the 107th Annual General Meeting, *Monthly Notices* (Feb. 1927), **87**, 247.) Thus did Jeans enable the Royal Astronomical Society to perpetuate the memory of Sir George Darwin, President of the Society for 1899–1900, who, as shown earlier in this volume, had befriended Jeans in his early years and had inspired, directly or indirectly, some of Jeans's most notable researches.

The following letter from Jeans to Hale explains itself.

Cleveland Lodge, Dorking
Dec. 11, 1926

Dear Hale,

We have heard with much pleasure that your health is reported to be much better than formerly. I hope most sincerely that the improvement is all that it is represented to be.

The report has emboldened the Royal Astron. Society to make a request which I, as President, am asked to transmit to you.

The Society has a new lectureship, 'The George Darwin' lectureship, to be given annually on any branch of astronomy whatever. Council are most anxious to give it a good start by getting a really eminent astronomer to give the first lecture, and to head the list of lecturers for all time. Discussion yesterday showed that a large number of members of Council had independently thought of you as the one proper and pre-eminently suitable first lecturer. No other name was mentioned in opposition to yours, and I was directed by unanimous vote to ask you if you could do us the great service of giving us the first 'George Darwin' lecture, if your health permits of your so doing.

Council ought formally to appoint the lecturer on Jan. 14th. Any day after this except Feb. 11th would be possible for the lecture. We had no information as to your plans, but if you are coming to this side, I am sure Council would be most anxious to arrange the day to suit your convenience. I shall hope to have your reply before Jan. 14th, and greatly hope it may be 'yes'.

Many thanks for your little book which I read with great admiration. I have been thinking over sun spot problems in connexion with differential rotation in the outer layers of the sun, but I don't see much light at present.

With best regards to Mrs Hale and yourself,

Yours sincerely,
J. H. JEANS

From the following letter it is evident that Hale's reply was 'No'.

Cleveland Lodge, Dorking
Jan. 19, 1927

Dear Hale,

It is good news that you are so much better, but I am very sorry the improvement is not enough to justify your undertaking the Lecture. But it is good to hear that you are so much more comfortable than you used to be, and I can quite appreciate that under the conditions you describe it would be very unwise to try to come and lecture—I am glad to have your suggestion as to Michelson. For the present we are now asking Schlesinger, as he is this year's medallist. I believe he's coming over here in any case, so we anticipate he will accept.

Many thanks for your kind remarks about visiting California. I really hope to do so before very long, and to have a less hurried visit than last time. At present I am strictly limited by R.S. necessities to Olivia's school times, but hope for better opportunities before long. I can say in confidence that I tried to get free from the R.S. last year: I sent in my resignation but they would not let me off, and now I feel I ought to stay on to the end of my ten years term of office—1929—or something approximative thereto.

All this administration and committee work makes an awful

hole in the time and energy available for research, as I expect you realize only too well. I have to go to London, for instance, for 5 days this week—about 15 committees and meetings. But I hope to find more time for myself before long, and one of the things I really look forward to is a leisurely visit to Mt Wilson.

When that time comes I much hope we may find you in Pasadena, and in still better health than now.

Again with regrets that you could not give us the pleasure of being our first George Darwin lecturer, but with hopes for the not remote future,

Yours sincerely,

J. H. JEANS

The first George Darwin Lecture was delivered on 11 March 1927, by Professor Frank Schlesinger, the gold medallist of the year. Its title was 'Astronomical Photography of Precision'. From that time on, the lecture has been regularly delivered by an astronomer from abroad, except on one occasion. The list of lecturers includes the names of the world's most distinguished astronomers during the past twenty years. The exceptional occasion was in 1944, when the lecture was delivered by Joseph Proudman, on the highly apposite subject, 'The Tides of the Atlantic Ocean' (*Monthly Notices* (1944), **104**, 244). The appositeness of this subject derives from the fact that, as mentioned by Proudman in opening his lecture, from about 1882 until the end of his life in 1912, Sir George Darwin was generally regarded as the greatest living authority on ocean tides.

Since, in accordance with Jeans's original suggestion, the George Darwin lecturer for the year has often been the Gold Medallist for the year, it has frequently happened that the subject of the lecture has been the original work for which the Gold Medal was awarded, and the President has often thought fit to deliver a shortened form of address on presenting the medal. This is perhaps to be regretted, but, on the other hand, the existence of the George Darwin Lecture will often set the President free to address the

Society at its annual meeting on some subject of his own choice and interest.

In any event Jeans's generosity in endowing the lectureship has very considerably added to the interest of each year's session of the Society.

In spite of the demands of the Secretaryship of the Royal Society, Jeans showed great scientific activity during the whole of the decade. In 1922 he delivered the Halley Lecture before the University of Oxford, his title being 'The Nebular Hypothesis and Modern Cosmogony'. The lecture, afterwards published by the Clarendon Press, opened with an account of Laplace's hypothesis as described by him in his *Système du Monde* (1796). It goes on to give an account of the evolution of rotating masses similar to that of the Adams Prize Essay of 1917, and concludes with the result, indicated, as Jeans is careful to insist, rather than proved, that whilst rotation can account for spindle-shaped nebulae, spiral nebulae, and double stars, it cannot account for a solar system of planets; that the formation of the solar system must rather be attributed to the disruption of the sun by tides raised by a passing star; that an encounter sufficiently close would be a very rare event; and that accordingly planetary systems like the solar system must be very rare in space.

But he who reads this lecture today should be warned that Jeans unfortunately made use of some supposed observational evidence which has not been confirmed and indeed has been contradicted by later researches. Since Jeans had published his Adams Prize Essay in 1919, there had appeared in the *Astrophysical Journal* a series of papers by the talented observer Adrian van Maanen, who had compared photographs of spiral nebulae taken at intervals of eleven years, and by measurement of the apparent changes appeared to be able to show that the four well-known spiral nebulae, Messier 51, 81, 101 and 33, were in states of rotation with periods of the order of only 100,000 years. Combination of this value with their spectrographically determined linear velocities of rotation gave their distances and sizes, and placed them well within the confines of our own galaxy. This is now known to be quite false: the periods of rotation of these spirals are of the same order of magnitude as the period of rotation of our own galaxy, which is of the order of at least 200,000,000 years. Van Maanen's measurements, in fact, carefully controlled as they were at the time, have never been confirmed, though it is not known how he came to be so misled. Those were the days before it was settled and accepted that the spirals are island universes, comparable with our own galaxy in size and stellar population; there was great controversy at one time on this matter. Jeans is not to be blamed for having accepted van Maanen's results at their face value. But the whole story is a warning to theoreticians, such as Jeans was, not to accept too easily the observations of the practical astronomer.

Jeans delivered the Rouse Ball Lecture before the University of Cambridge in 1925, on Atomicity and Quanta. This is a brilliant exposition, in sixty-four pages (Cambridge University Press, 1926), of the then state of the quantum theory of the atom and of radiation. Though Jeans had scarcely written any technical papers on atomic physics since his famous *Report* of 1914, this lecture shows that, in spite of his own preoccupation with cosmogony and his arduous duties as Secretary of the Royal Society, he had followed the development of the quantum theory with indefatigable interest. A few quotations from the lecture may be of interest. Speaking of the Greeks, he said (p. 7):

Granted, however, that it was natural, and perhaps inevitable, that space should be deemed continuous and matter atomic, what is to be said about time? Clerk Maxwell makes the statement, I know not on what authority, that up to the time of Zeno 'time was still regarded as made up of a finite number of moments, while space was confessed to be divisible without limit'. Jowett, it is true, attributes to Plato the statement that 'time is to arithmetic what space is to geometry', but this appears to mean nothing more recondite than that time is one-dimensional, while space is three-dimensional. If the Greeks really regarded time as atomic—as a succession of moments—we have a simple explanation of the otherwise incomprehensible puerility of the famous paradoxes of Zeno. The best way of attributing any reasonable interpretation to these is perhaps to regard them as showing that the hypothesis of non-atomic space and atomic time lead to absurdities contrary to experience.

These remarks by Jeans are of special interest to me, since, in collaboration with G. J. Whitrow I have devoted much thought and energy to this very process of the *arithmetization* of physics by reducing all length measures and measures derived therefrom to time measures. From this point of view, the statement attributed to Plato by Jowett becomes much more recondite than was allowed by Jeans. But Jeans, as we shall see later, tended to be hostile to Greek philosophy.

Jeans's forcible way of arguing may be illustrated from this lecture by his account of Einstein's photochemical law that in a photochemical reaction the number of molecules which are affected is equal to the number of quanta of light which are absorbed. Jeans commented on this:

As a corollary to this law, the work done by one powerful quantum cannot be done by two weak quanta, or indeed by any number of weak quanta. If a quantum of a certain specified energy is needed to affect a molecule, quanta of lower energy, no matter how great their number, will produce no effect at all. The law not only prohibits the killing of two birds by one stone, but also the killing of one bird with two stones. Since the energy of a quantum h is proportional to its frequency, we understand how a small amount of violet light can accomplish what no amount of red light would suffice to do, a fact familiar to every photographer—we can admit quite a lot of light if only it is red, but the smallest amount of violet light spoils our plates.

A similar illustration is provided by the photoelectric effect. Quanta more energetic than those requisite for photochemical action may suffice to break up the atoms on which they act, giving rise to the photoelectric effect. Again we find that there is a limiting frequency of radiation below which action does not take place no matter how intense the light. But as soon as the frequency passes this limit, even the feeblest intensity of light starts photoelectric action at once. Again it is found that the absorption of one quantum breaks up the atom, or rather ejects one electron from the atom; if the energy of the quantum is greater than that necessary to remove the electron from the atom, the excess is used in setting the electron into motion.

In this way we begin to see that what is atomic is not material energy at all, as Planck originally thought, but radiant energy, or to be more precise the exchanges of energy between radiation and matter. We begin to understand why the ultraviolet end of the spectrum of a hot body shows so little energy; the simple reason is that very, very few of the molecules or atoms in a hot body possess enough energy to emit a complete quantum of violet radiation. In a hot body at temperature T, the vast majority possess energy of amount comparable with RT, where R is the absolute gas-constant. The proportion of atoms whose energy is a large multiple of this, say nRT, where n is a large number, is of the order of e^{-n} (or strictly, its logarithm to base e is of the order $-n$). Thus, when n is large, this proportion is very small indeed. If the quantum of radiation of any frequency ν is equal to nRT, where n is large, the proportion of atoms which are in a position to shed a whole quantum $h\nu$ of radiation is a small quantity of the order of e^{-n} or $e^{-h\nu/RT}$, whence it follows that the amount of radiation of high frequency ν is also a small quantity of the order of $e^{-h\nu/RT}$....

Planck's fundamental discovery can be stated in the form that h is different from zero.... Small though the value of h is, we must recognize that it is responsible for keeping the universe alive. If h were strictly zero, the whole material energy of the universe would disappear into radiation in a time which would be measured in thousand-millionths of a second.

He concluded this lecture on a pessimistic note:

But in nature it is only for radiations of very low frequency that any period in the atom coincides with that of the radiation which it emits or absorbs. We may think of the atom as acting as a 'source' for radiation of certain definite frequencies, and as a 'sink' for radiations of other definite frequencies, but as a rule the atom does not beat time to any of these frequencies. Only in the extreme case of excessive slowness do the 'source' and 'sink' actions reduce to ordinary emission and absorption by resonance; in the more general case, I know of no phenomenon in the whole of physics which helps us in the least to comprehend the physical processes at work. With this fact before us, not much meditation is needed to convince us that we are still very far from understanding the working of the atom or the true meaning of atomicity and quanta.

This pessimistic ending is unusual and it is difficult to see what Jeans meant by talking of 'sources' of radiations of certain definite frequencies and 'sinks' for others; this would seem to contradict Kirchhoff's law of emission and absorption.

But acting as Secretary of the Royal Society, presiding over the Royal Astronomical Society and delivering important named lectures did not comprise the whole of Jeans's activities during this decade. From his first paper in the *Monthly Notices* in 1913 to his last paper in the same journal of 1928, Jeans made some thirty-five contributions to astronomy. It is not necessary at this point to consider them in detail, for he himself welded them together in his large technical volume, *Astronomy and Cosmogony* (1928), which will be discussed in Chapter XI. Here it may be remarked that this work scarcely ranks with the Adams Prize Essay, *Problems of Cosmogony and Stellar Dynamics* of 1919, to which it is in some sense a sequel, though it is planned on a more comprehensive scale. But more ambitious in aim as it is, it scarcely competes with the earlier work in clarity or permanence. The earlier work compelled attention by the force and vigour and even majesty of its mathematics; but in the work of 1928, some of the mathematics and a good deal of the physics are far from compelling, and it is not written with the sure physical grasp that we should have expected from one of Jeans's rank. Also, its main theses are far from being accepted, either now or at the time it was written. It has never been cited to the same extent as Eddington's *Internal Constitution of the Stars*, though I personally believe that in spite of the accuracy of Eddington's physical detail, the underlying conceptions of Eddington's attack are false, whilst in spite of the unphysical nature of Jeans's assumptions in *Astronomy and Cosmogony*, his underlying ideas were sounder than Eddington's. Jeans's Adams Prize Essay of 1919 was and remains a classic, even where subsequent discoveries have proved it wrong. *Astronomy and Cosmogony* cannot claim to be a classic; nevertheless, it makes fascinating reading from beginning to end. It is stimulating even where it irritates, and instructive even where it mystifies. It is usually more profitable to be irritated by Jeans than lulled by lesser writers. *Astronomy and Cosmogony* exhibits the subject of stellar structure as in a state of flux, when nebular recession and nebular rotation were less well established than today, and when the confusion of the time-scales was unresolved. But in such a subject as cosmogony it is perhaps improper ever to expect finality or consensus of opinion, and *Astronomy and Cosmogony* certainly served a real and valuable purpose in gathering together a great mass of material and attempting its consolidation.

I have devoted this and most of the preceding chapters to Jeans's researches and public activities, but a few words may be added about his private life. After his retirement from the Stokes Lectureship at Cambridge in 1912 he went to live first at Guildford and then in London. He spent the summer of 1913 on Dartmoor, returned to London in 1914, lived for the summer of 1914 at Amersham and spent the summer of 1915 at Brighton. He spent the summer of 1916 at Box Hill, the summer of 1917 at Holmbury St Mary. In 1918 he acquired Cleveland Lodge, Dorking, and settled there, taking a great interest in the garden, though he was no gardener himself. He loved to stroll round the grounds of Cleveland Lodge, or sit quietly on the lawn. He had an organ installed in the house and often played three or four hours a day, but purely for his own satisfaction, never in front of other people.

The first signs of heart disease attacked him in 1917, but he appeared at the time to shake this off.

CHAPTER V

POPULAR EXPOSITION
1929–30

JEANS was knighted in 1928 for his services to science and to the Royal Society, and it is noteworthy that this honour came to him before the publication of any of his popular books. To this phase of Jeans's life we now come.

Astronomy and Cosmogony (1928) concluded with a very moving chapter, in which Jeans summed up, without mathematics but with some vivid diagrams, his life-work of research in the 'natural history' of the astronomical formations—galaxies, stellar clusters, nebulae, stars (simple, double and multiple), Cepheids, novae and solar systems—which appear to constitute the material universe. The three concluding paragraphs of this chapter may be quoted *in extenso*:

The cosmogonist has finished his task when he has described to the best of his ability the inevitable sequence of changes which constitute the history of the material universe. But the picture which he draws opens questions of the widest interest not only to science, but also to humanity. What is the significance of the vast processes it portrays? what is the meaning, if any there be which is intelligible to us, of the vast accumulations of matter which appear, on our present interpretations of space and time, to have been created in order that they may destroy themselves? What is the relation of life to that universe of which, if we are right, it can occupy only so small a corner? What if any is our relation to the remote nebulas, for surely there must be some more direct contact than that light can travel between them and us in a hundred million years? Do their colossal incomprehending masses come nearer to representing the main ultimate reality of the universe, or do we? Are we merely part of the same picture as they, or is it possible that we are part of the artist? Are they perchance only a dream, while we are brain-cells in the mind of the dreamer? Or is our importance measured solely by the fractions of space and time we occupy—space infinitely less than a speck of dust in a vast city, and time less than one tick of a clock which has endured for ages and will tick on for ages yet to come?

It is not for the cosmogonist to attempt to suggest answers to these wide questions, or even to the more limited questions directly raised by the sequence of events which is his own special study. He will be specially reluctant to attempt either, knowing how dimly most of the sequence of events can be seen, and how much of it cannot be seen at all. His critics may allege that what he sees most clearly is only a creation of his own imagination, and he is only too conscious that it may be so. He can only end with a question; others, more confident or more fortunate, may, if they wish, attempt an answer.

Let us, however, reflect that mankind is at the very beginning of its existence; on the astronomical time-scale it has lived only for a few brief moments, and has only just begun to notice the cosmos outside itself. It is, perhaps, hardly likely to interpret its surroundings aright in the first few moments its eyes are open.

This chapter attracted the notice of R. H. Fowler who remarked upon it to S. C. Roberts, the Secretary of the University Press. Roberts had already been reminded by a colleague that the quality of Jeans's style had gained him a place in *The Oxford Book of English Prose* and he approached Jeans about a popular book.*

Thus it came about that Jeans gave up technical work, almost overnight as it were, on the completion of *Astronomy and Cosmogony*, and began to devote himself to exposition. *Astronomy and Cosmogony* appeared in 1928, and his last technical original paper, entitled 'Liquid stars, a correction', appeared in the *Monthly Notices* for 1928. In 1929 appeared *The Universe Around Us*, his first popular book, a volume of 352 pages. The book was partly based on a series of

* See Memoir.

broadcast talks given in the autumn of 1928, partly on other lectures, but it was a complete rewriting of the original texts.

As Jeans says in his preface, he gave special attention to problems of cosmogony and evolution, and to the general structure of the universe. His initial chapter headings, 'Exploring the Sky', 'Exploring the Atom', 'Exploring in Time', describe respectively the astronomical, physical and cosmological interests dealt with, and the concluding chapter, 'Beginnings and Endings' dwelt on the results Jeans had himself obtained in his wide researches on the evolution of celestial bodies and astronomical eschatology. Like his technical treatises, this book sustains the reader's excited interest from cover to cover. Few astronomers, I suppose, accepted all its conclusions, even at the time of publication; still less would they do so today when the expansion of the universe is the dominating feature of the universe of spiral nebulae, a phenomenon which had hardly begun to be realized at the time Jeans wrote. But cosmogony and cosmology are subjects of which the complete tale cannot even yet be told, and any attempt to tell the whole tale must necessarily have weak spots in it. Jeans's view that the universe began as one huge, widely extended, all-space-filling gaseous nebula is not compatible with our own present knowledge, which indicates rather that the universe began as a complicated point-singularity. Nevertheless, the large synthesis that Jeans effected was well worth doing. The success of the book was world-wide and immediate, and thoroughly well-deserved.

The next year, 1930, Jeans delivered the Rede Lecture before the University of Cambridge, and an expansion of this lecture, under the title *The Mysterious Universe*, was immediately published, and had an immense circulation. It was regarded by Jeans as a sequel to *The Universe Around Us*, but it was written so as to be complete in itself. The five chapters in it (it amounted to 150 pages) were entitled 'The Dying Sun', 'The New World of Modern Physics', 'Matter and Radiation', 'Relativity and the Ether' and 'Into the Deep Waters'.

It was the last of these chapters that attracted the most attention and, indeed, created a sensation in the Press the morning after the lecture. In it Jeans came to his famous conclusion that the Great Architect of the Universe must be a mathematician: 'We have already considered with disfavour the possibility of the universe having been planned by a biologist or an engineer; from the intrinsic evidence of his creation, the Great Architect of the Universe now begins to appear as a pure mathematician.'

Let me trace the argument of this chapter, 'Into the Deep Waters', by quoting the salient sentences. In the previous chapter, devoted to relativity, Jeans had adopted the view that 'a soap-bubble, with irregularities and corrugations on its surface, is perhaps the best representation...of the new universe revealed to us by the theory of relativity'. He discerned amongst these irregularities and corrugations two main kinds, interpretable as matter and radiation, the two ingredients of the universe.

We may think of the surface of the bubble as a tapestry whose threads are the world lines of atoms....As we move timewards along the tapestry, its various threads for ever shift about in space and so change their places relative to one another. The loom has been set so that they are compelled to do this according to definite rules which we call 'laws of nature'....

Your consciousness touches the picture only along your world line, mine along my world line, and so on. The effect produced by this contact is primarily one of the passage of time....It may be that time, from its beginning to the end of eternity, is spread before us in the picture, but we are in contact with only one instant, just as the bicycle-wheel is in contact with only one point of the road. Then, as Weyl puts it, events do not happen; we merely come across them....

Jeans proceeds to quote Plato in the *Timaeus* to a similar

effect. He shows that the waves which were at one time supposed to traverse the 'ether' have been reduced to little more than an abstraction.

This quality of abstractedness in what were at one time regarded as material 'ether-waves' recurs in a far more acute form when we turn to the system of waves which make up an electron. The 'ether' in terms of which we find it convenient to explain ordinary radiation...has three dimensions in space in addition to its one dimension of time. So also has the ether in which we describe a single electron isolated in space, (though) this may not be the same ether as before. But a single electron isolated in space provides a perfectly eventless universe, the simplest conceivable event occurring when two electrons meet one another. To describe (this)..., wave-mechanics asks for a system of waves in an ether which is of seven dimensions.... Most physicists would agree that the seven-dimensional space... of two electrons is purely fictitious, in which case the waves which accompany the electrons must also be regarded as fictitious....Yet it is hard to see how one can attribute a lower degree of reality to the one set of waves than to the other: it is absurd to say that the waves of one electron are real, while those of two electrons are fictitious.

He goes on to mention that the electron waves have been regarded as waves of probability, which, in accordance with Heisenberg's 'uncertainty principle', 'makes it impossible ever to say: an electron is just here'. He quotes Dirac as extending this indeterminacy and uncertainty of knowledge over the whole of atomic physics, and Heisenberg and Bohr as suggesting that electron waves must be regarded merely as a sort of symbolic representation of our knowledge as to the probable state and position of an electron. 'If so, they change as our knowledge changes, and so become largely subjective. Thus we need hardly think of the waves as being located in space and time at all; they are mere visualizations of a mathematical formula of an undulatory, but purely abstract, nature.' We may have to go even further, and deny the possibility of atomic phenomena occurring in space and time. To himself (he confesses) this seemed the most promising interpretation of the situation.

> Just as the shadows on a wall form the projection of a three-dimensional reality into two dimensions [he has quoted Plato's cave-shadows earlier] so the phenomena of the space-time continuum are four-dimensional projections of realities which occupy more than four dimensions, so that events in time and space become
>
> No other than a moving row
> Of Magic Shadow-shapes that come and go.
>
> ...The essential fact is simply that *all* the pictures which science now draws of nature, and which alone seem capable of according with observational facts, are *mathematical* pictures.

Science, he says, finds in a whole torrent of surprising new knowledge that the way which explains them more clearly, more fully and more naturally than any other is the mathematical way, the explanation in terms of mathematical concept.

> So true is it that no one except a mathematician need ever hope fully to understand those branches of science which try to unravel the fundamental nature of the universe—the theory of relativity, the theory of quanta, and the wave-mechanics.... Nature seems very conversant with the rules of pure mathematics, as our mathematicians have formulated them in their studies, out of their own inner consciousness, and without drawing to any appreciable extent on their experience of the outer world. By 'pure mathematics' is meant those departments of mathematics which are creations of pure thought, of reason operating within her own sphere.

He then discusses the attempt to discover the nature of the reality behind the shadows. All such discussion must be barren unless we have extraneous standards against which to compare them; hence, in Locke's phrase, 'the real essence of substances' is for ever unknowable. Only the laws which govern the changes of substances are capable of discussion and of comparison with the abstract creations of our own minds. These laws suggest the conclusion that, put very crudely and inadequately, the universe appears to have been designed by a pure mathematician.

Jeans foresees that his conclusion may be challenged on two grounds. First, there is a chance that we are merely moulding nature to our preconceived ideas, much as a musician wholly engrossed with music might regard every piece of mechanism as a musical instrument, or as a cubist painter might reduce all nature to piles of cubes. But Jeans denies that this can be the whole story, since we may be warned by two former failures to interpret nature: the failure of our remoter ancestors to interpret nature in terms of anthropomorphic concepts, and the failure of our immediate ancestors to achieve an interpretation in terms of engineering concepts—models of billiard balls, strings, springs and the like. As against this, Jeans claims that the attempt to interpret nature in terms of mathematical concepts has been brilliantly successful. Secondly, Jeans says his conclusion may be challenged on the ground that there is absolutely no sharp line of demarcation between pure and applied mathematics. He admits that some of the concepts with which the pure mathematician works are taken direct from his experience of nature.

If, however, the more intricate concepts of pure mathematics have been transplanted from the workings of nature, they must have been buried very deep indeed in our subconscious minds. This very controversial possibility is one which cannot entirely be discussed, but at any event it can hardly be disputed that nature and our conscious mathematical minds work according to the same laws. She does not model her behaviour, so to speak, on that forced on us by our whims and passions, or on that of our muscles and joints, but on that of our thinking minds. This remains true whether our minds impress their laws on nature, or she impresses her laws on us, and provides a sufficient justification for thinking of the designer of the universe as a mathematician.

Jeans carries the idea even further. The material in which the pure mathematician works is pure thought: his creations are not only thought out, but consist of thought—just as the creations of an engineer consist of engines. The concepts required for our understanding of nature—finite, expanding space, spaces of many dimensions, probability laws instead of causation laws—all these and others Jeans feels bound to regard as structures of pure thought. 'If all this is so, then the universe can be best pictured, although still very imperfectly and inadequately, as consisting of pure thought, the thought of what, for want of a wider word, we must describe as a mathematical thinker.'

So he concludes that modern science arrives, if by a very different path, at the same Idealism that Bishop Berkeley arrived at:

in the stately and sonorous diction of a bygone age.

All the choir of heaven and furniture of earth, in a word all those bodies which compose the mighty frame of the world, have not any substance without the mind.

The 'objectivity [of objects] arises from their subsisting "in the mind of some Eternal Spirit".'

But he will not discard Realism altogether: it is merely that the boundary between Realism and Idealism has become very blurred: or rather the boundary is a relic of a past age, when reality was conceived as identical with mechanism. He prefers the label 'mathematical' to either of the labels 'real' or 'ideal', provided the adjective 'mathematical' is held to connote something beyond the studies of the professional mathematician. This does not detract from the substantiality of things, or the existence of external reality. He quotes a telling phrase of Chalmers Mitchell, 'the element of surprise is sufficient warrant for external reality', a second warrant being 'permanence with change—permanence in your own memory, change in externality'. Jeans admits these as arguments against solipsism, but belittles the argument from surprise, as 'powerless against the concept of a universal mind of which your mind and mine, the mind which surprises and that which is surprised, are units or even excrescences'.

He distinguishes between degrees of substantiality, contrasting the space of a dream with the space of everyday life; he tries to make a similar distinction regarding time, but in speaking of 'the time of the universal mind' he lapses for a moment into pre-relativity grooves of thought. However, he claims that the concept of the universe as a world of pure thought throws light on several situations he has encountered, in earlier portions of the lecture, in his survey of nature: for example, how the 'ether' is a mere abstraction; why energy, 'the fundamental entity of the universe', is just the constant of integration of a differential equation —forgetting that Einstein had relegated it to the status of a mere component of a four-dimensional tensor.* But there is much in Jeans's contention that the truth about a phenomenon resides and resides only in its mathematical description. 'We go beyond the mathematical formula at our own risk'—a risk, however, Jeans never tired of taking. He admits that a model or picture may help towards understanding, but we have no right to expect a model or picture to be possible. In one of his finest passages, he develops this theme.

The making of models or pictures to explain mathematical formulae and the phenomena they describe is not a step towards, but a step away from, reality; it is like making graven images of a spirit. And it is as unreasonable to expect these various models to be consistent with one another as it would be to expect all the statues of Hermes, representing the god in all his varied activities—as messenger, herald, musician, thief and so on—to look alike. Some say that Hermes is the wind; if so, all his attributes are wrapped up in his mathematical description, which is neither more nor less than the equation of motion of

* Kinematic relativity rehabilitates energy, showing that it is both a constant and a scalar invariant.

a compressible fluid. The mathematician will know how to pick out the different aspects of the equation which represent the conveying and announcing of messages, the creation of musical tones, the blowing away of our papers, and so forth. He will hardly need statues of Hermes to remind him of them, although, if he is to rely on statues, nothing less than a whole row, all different will suffice.

And then, in a somewhat caustic sentence: 'all the same, the mathematical physicist is still busily at work making graven images of the concepts of wave-mechanics.'

On the view that the essential nature of the universe is mathematical, he says

We need find no mystery in the nature of the rolling contact of our consciousness with the empty soap-bubble we call space-time, for it reduces merely to a contact between mind and a creation of mind—like the reading of a book, or listening to music. It is probably unnecessary to add that, on this view of things, the apparent vastness and emptiness of the universe, and our own insignificant size therein, need cause us neither bewilderment nor concern. We are not terrified by the sizes of the structures which our own thoughts create, nor by those that others imagine and describe to us.

This last argument seems to me unfair; the vastness of the universe, and man's insignificance therein, are data of observation, not of imagination; for Jeans cannot have it both ways, he cannot admit substantiality and then later appeal to a false unsubstantiality. But he ends this argument with a human touch—human, that is, for an astronomer: 'the immensity of the universe becomes a matter of satisfaction rather than awe; we are citizens of no mean city'.

He lulls us into thinking that there is no longer anything to puzzle about in the finiteness of space. He is on surer ground in finding indications that time also is finite, or at least in its backward extent. There must be a 'time' before which the present universe did not exist. (We call this nowadays the 'origin of time', and regard it as the epoch of a mathematical singularity in density.) But Jeans still considers that thermodynamics teaches the doctrine of the 'heat-death', the doctrine that everything in nature passes to its final state by a process we call the 'increase of entropy'. Broadly speaking, this means that in a finite space containing a finite amount of matter, temperature must tend to equalize itself, until a state is reached in which no further change is possible, in which entropy has reached its maximum. If this occurs, the universe will become dead. This process is irreversible. Jeans uses this as another argument in favour of a transcendental act which we must call the 'creation' of the universe, at a time not infinitely remote.

I accept the argument for a beginning of things, a finite time ago; we must be careful, however, to say to what scale of time we are referring. The full meaning of this I have worked out in writings elsewhere. Here it is interesting to note the acuteness of Jeans's insight into the actual state of the case. But I do not accept the argument for the 'heat-death' of the universe as its ultimate and therefore changeless state. The argument that since the entropy of any limited portion of the universe is increasing, it must once have begun at a low value—that, if the universe is like an unwinding watch, it must at one time have been wound up—is at first sight attractive; but we have now to take into account the phenomenon of the expansion of the universe, in the scale of time in which the universe had a beginning a finite time ago. In the model of the universe to which kinematic relativity is led, the singularity representing 'creation' is to be found again, in our world-wide present instant, on the frontiers of the finite volume of expanding space which the universe occupies; this means that the finite space contains an infinite amount of matter, the overwhelming proportion of it concentrated towards the inaccessible boundary. But when the universe, though occupying a finite space, contains an infinite amount of matter, the argument for the 'heat-death' fails: there are always, near the frontier, youthful, newly-created galaxies, and the world has a perennially young frontier. The second law of thermodynamics applies only to enclosed systems, whilst the universe, though occupying a finite space, is essentially an unenclosed system, an 'open' set of particles.

It must be said that this view of the universe was developed only after Jeans had written *The Mysterious Universe*. The only criticism that can be legitimately made of Jeans's argument here is that whilst it was valid in the then state of knowledge and theory regarding cosmology, it was stated too positively. Jeans did indeed add the proviso 'unless this whole branch of science is wrong'. But it is not a question of the whole of this branch of science being wrong; it is that the actual state of the universe probably corresponds to exceptional features, which allow of exceptions to the 'heat-death' inference.

But Jeans had not finished with the drawing of inferences. He went on to say that

> if the universe is a universe of thought, then its creation must have been an act of thought....Time and space must have come into being as part of this act. Primitive cosmologies pictured a creator working in space and time, forging sun, moon and stars out of already existent raw material. Modern scientific theory compels us to think of the Creator as working outside time and space, which are part of his creation, just as the artist is outside his canvas.

And he quotes Plato to the same effect.

He insists that scientific explanations of things, to be convincing, must be simple. And while he admits that the mathematical explanation may prove neither to be final nor the simplest possible, he maintains that

> we can unhesitatingly say that it is the simplest and most complete so far found, so that, relative to our present knowledge, it has the greatest chance of being the explanation which lies nearest to the truth....Mechanics has already shot its bolt, and

failed dismally, on both the scientific and the philosophical side. If anything is destined to replace mathematics, there would seem to be specially long odds against its being mechanics.

As against the view that into a 'fortuitous concourse of atoms' life had stumbled by accident, and, with it, mind, the universe now begins to look more like a great thought than a great machine, with mind no longer an intruder but rather the creator and governor of the realm of matter—not our individual minds but the mind in which the atoms, out of which our individual minds have grown, exist as thoughts.

Again, he argues that the new knowledge provided by modern science compels us to revise any hasty first impressions that

we had stumbled into a universe that either did not concern itself with life or was actively hostile to life. The old dualism of mind and matter...seems likely to disappear, not through matter becoming more shadowy or insubstantial...or through mind becoming resolved into a function of the working of matter, but through substantial matter resolving itself into a creation and manifestation of mind.

He repeats that 'the universe shows evidence of a designing or controlling power that has something in common with our own individual minds—not, so far as we have discovered, emotion, morality or aesthetic appreciation, but the tending to think in the way which, for want of a better word, we describe as mathematical'. Thus Jeans's mathematical God is denied any connexion with the good or the beautiful, with ethics or aesthetics. If the standard values of the good, the true and the beautiful correspond to ethics, science, and aesthetics, it is not surprising that value approached solely from the scientific side should only disclose what for Jeans's contemporaries was the then truth, and that God as thus approached should be only a God of order.

The physicists of any one epoch, it has been well observed, agree with one another, as a rule, in their conclusions; but these conclusions disagree with those of the physicists of twenty years before, and with those of twenty years later—at least in this century. The philosophers of any one epoch disagree violently with one another, but over long periods of time they agree in their formulations of problems, though not in their conclusions (it is scarcely the job of the metaphysician to reach conclusions). How are we to assess the permanence of the conclusions of the physicist turned philosopher? Jeans was fully aware that this question would be raised, and he ends his lecture with a warning as to the provisional nature of his conclusions.

> We may well conclude by adding, what might well have been interlined into every paragraph, that everything that has been said, and every conclusion that has been tentatively put forward is quite frankly speculative and uncertain. We have tried to discuss whether present-day science has anything to say on certain difficult questions, which are perhaps set for ever beyond the reach of human understanding....Perhaps...science should leave off making pronouncements: the river of knowledge has too often turned back on itself.

I have thought it worth while to sketch, largely in Jeans's own words, the content of the concluding chapter of the Rede Lecture of November 1930, so that the reader may be in a position to understand as far as possible why this book had such a remarkable success. I have also introduced some criticisms, which could scarcely have been made at the time the lecture was delivered. The periodical press was filled with correspondence on the subject of science and religion, God and man, time and eternity, but in spite of those who argued that Jeans was conceiving God in his own image, the deep sincerity of Jeans's lecture appealed to everyone. Jeans did not expect people in general to agree with him; he said so in his preface; he said, moreover, that it was his intention to provoke disagreement. But he can scarcely have foreseen the enormous circulation the lecture would have.

This must have influenced Jeans towards employing his gifts for exposition. Certain it is that from now on he did no more original work of a technical kind. Many of his fellow-scientists were shocked at this, or appeared to be so. But I think Jeans was right. His constructive work was, perhaps, finished. He had great scientific achievements to his credit; in 1929 he was fifty-two; and nothing is more pathetic in the biography of a great scientist than when the stream of original papers, instead of drying up suddenly, becomes a trickle of inferior quality. Jeans had probably passed the height of his powers as a mathematician. Probably he recognized this.* The less he concerned himself now with the technical details of mathematical investigation, the more he could stand back and survey what he himself and others had accomplished, and interpret it for the benefit of the intelligent non-specialist. His decade of astronomical activity, 1919–29 had been splendid, but scarcely the equal of the half-decade, 1914–19. Instead of forcing himself into the narrower grooves of technical original work again, he embarked on a second career, which was to be at least equally splendid. Many of those who criticized him could probably conduct original investigations which would surpass any that were likely to come from Jeans; but hardly one of them could come near him in felicity of English, in clarity of exposition, in aptness of simile, or in width of knowledge. No one has the right to blame him for what must have been a difficult, though deliberate, decision.

* Lady Jeans has since told me that he was conscious that his powers as a mathematician (like those of so many mathematicians) declined in later years.

CHAPTER VI

LATER YEARS, 1931–46

MANY honours came to Jeans. The Merchant Taylors' Company admitted him to the Honorary Freedom of the Company—a rare distinction—and he received honorary degrees from many universities, including those of Oxford, Manchester, Dublin, Benares, St Andrews, Aberdeen, Johns Hopkins and Calcutta; but no award gave him greater pleasure than that of the Franklin Medal by the Franklin Institute of Philadelphia in 1931. On 24 February of that year he wrote to G. E. Hale:

At last, to my great pleasure, I find it is possible to visit Mount Wilson, as far as I can tell in the first fortnight in May. I am writing at once to enquire whether there is any prospect of seeing you at that time in Pasadena, or if you will not be there, where you are likely to be. You have probably seen in the newspapers that the Franklin Institute have been good enough to award me their Medal, and I am crossing to receive it on May 20th. I shall leave here as soon as Olivia returns to College, which I think is the 17th April, and shall come almost directly to the Observatory.

I much hope it may be possible to see you somehow or other. My wife joins me in sincerest good wishes to Mrs Hale and yourself. We both hope your health is much better.

Hale, in reply, expressed his delight that the Franklin Institute had voted their highest distinction to Jeans and went on:

Adams and Millikan, who are arranging for the coming meeting of the American Association for the Advancement of Science in Pasadena, have cabled to ask if you can come after instead of before the Franklin Institute presentation, so as to be the principal speaker during the sessions, which extend from June 15 to June 20. Although I still have to avoid all scientific and social

Dr Walter S. Adams, Sir James Jeans, and Mr Edwin P. Hubble at Mount Wilson observatory

functions (thereby missing the many opportunities afforded by Einstein's visit), I sincerely hope you can accept this invitation, as they naturally wish to make this first meeting of the whole Association in the West as successful as possible. Another advantage would be that you could make us a longer visit. Moreover, several of the leading men of the California Institute and the Observatory will be away in May, attending the meetings in the East of the National Academy, American Philosophical Society and other Societies.

As for myself, I may possibly find it necessary to go East in May, though not to attend any meetings. In any case, I would arrange to see you somewhere, as I must not lose such a rare and welcome opportunity. I had a considerable operation last autumn, and while it has temporarily relieved, though not removed entirely, an old local trouble (one of the causes of my immobility in London), it seems to have left me as weak as ever in the upper storey. The result is that I am still forced to live a hermit's life, doing a small amount of observing and writing here in my laboratory, but compelled, greatly against my will, to avoid the Observatory and the laboratories of the Institute, and to see only two or three people here a week, because of the exhausting effects of interesting talks of any kind. But this is better than the complete seclusion I have been compelled to undergo for long periods in the past after more severe but less chronic nervous breaks.

I was very sorry indeed to hear of Sir Charles Parsons' death—a serious loss to science as well as to engineering.*

Mrs Hale joins me in warmest regards to you all and in the hope that Lady Jeans may come with you. I wish also we might see Olivia, whom I remember with affection.

Jeans was obliged, however, to stick to his plan of an earlier visit to Pasadena and wrote on 16 April:

I was sorry I was unable to do as you, Adams and Millikan suggested and visit Pasadena in June for the meeting of the American Association. The difficulty is that I am not altogether a free man. Olivia is now at College at Cambridge and as my

* I remember hearing Rutherford describe Hale himself as a 'great engineer-astronomer'.

wife is coming with me to California, we have to fit our trip in within the limits of a College term, which ends in the middle of June. This made the latter date quite impossible. If I had known that the American Association was meeting in Pasadena at that time, I might have tried to make other arrangements for Olivia, but it was too late to do this when the invitation reached me.

I am so glad to hear that your health is, comparatively speaking, better than it was, but still hope to hear much better news of you before long. If I see you when I am in the States, I will be careful not to be too exacting, as I know it is still necessary to avoid all discussion of exciting topics with you. I am most desperately sorry that this is still the case, and can only hope for better news before long.

This will hardly reach you before the time of my arrival in Pasadena. I expect to be there on April 29th or thereabouts.

The presentation of the medal took place at Philadelphia on 20 May and Jeans delivered an address on 'The Origin of the Solar System'.

One result of the outstanding success of *The Universe Around Us* and *The Mysterious Universe* was that Jeans was in constant demand as a lecturer, both to learned societies and to popular audiences, and also as a broadcaster. *The Stars in Their Courses* (1931) was based upon wireless talks and *Through Space and Time* (1934) upon the Christmas Lectures given at the Royal Institution in 1933. Between these came *The New Background of Science* (1933) in which Jeans tried to exhibit the new knowledge in such a way that every reader could form his own judgement on its philosophical implications. 'I have not', he wrote, 'suppressed my own view that the final direction of change will probably be away from the materialism and strict determinism which characterized nineteenth-century physics, towards something which will accord better with our everyday experience.'

This rapid succession of popular and semi-popular works of consistently high literary quality provoked the astonishment both of the learned and of the unlearned. Jeans was dubbed 'the Edgar Wallace of Astronomy' and it was a Fleet Street joke to ask: 'Have you read the midday Jeans?'

Nor was it confined to one country. In the United States the books were sold in large numbers and they were translated into the languages of the world. *The Mysterious Universe*, in particular, appeared in French, German, Italian, Dutch, Danish, Norwegian, Portuguese, Polish, Swedish, Finnish, Czech, Bengali, and Burmese.

In 1934 Jeans's life was temporarily shattered by the death of his wife. He had taken her to Florida in the spring, but there her health grew worse. Jeans wrote of what he felt to his old friend G. E. Hale:

'I have been meaning to write to you for a very long time—indeed, I have not answered your kind note and invitation which came at Christmas time.

My whole life has been changed by the death of my dear wife which occurred at the end of May. She had been suspiciously ill for some time, but we went to Florida in March, and if she had proved to be well I had hoped we might cross over to Pasadena then, or possibly come again in the autumn. But in Florida she got actually worse and we hurried home only for her to die after a few weeks. Happily she passed very peacefully after only a few days of acute illness. It has been a terrible shock to me as we had been such good friends and inseparable companions—especially as she got more deaf and was reluctant to see friends or strangers.

I still hope I may get to California before long and hope to take up my research more actively than has been possible in the last few years. At present Olivia is living with me but she is temporarily out of action with a strained heart.

We have had Hubble over here as you know, but have been able to see very little of him. It was good to hear from him that there is real improvement in your health. May it continue to improve.

My warmest regards to Mrs Hale and yourself.'

For some time Jeans was borne down by his sense of loss. But gradually he recovered his interests and when Sir

William Hardy, President-Elect of the British Association, died suddenly before assuming office in the autumn of 1934, Jeans gallantly stepped into his place and delivered the presidential address at Aberdeen in his customary brisk and fluent style.

In the following year the Royal Institution decided to establish a professorship of astronomy and to invite Jeans to be the first holder of the chair. This post he held until 1946, when failing health compelled him to resign. He was succeeded by the Astronomer Royal, Sir Harold Spencer Jones.

But the year 1935 was destined to bring Jeans something more important than a professorship. In the summer of that year Susanne Hock, daughter of Oskar and Katharina Hock of Vienna and a brilliant organist with an international reputation at the age of twenty-four, came with Lady Heath to a musical party at Jeans's house and played the organ for two hours after dinner. Their second meeting was in Switzerland* and their mutual attraction was rapidly developed. They were married in Vienna in September after a week's engagement and the marriage brought immeasurable happiness to both of them.

At Cleveland Lodge there was already an organ which Jeans had built for his own use. But now a new organ-room was added and a new organ built for Susi, a 'baroque' organ in the seventeenth-century manner. The two organ-rooms were adjacent but were acoustically insulated, so that either organ could be played without disturbing the other.

Out of their musical interests came the book *Science and Music* (1938). The idea was Susi's, and to Susi it was dedicated. The book, which contains chapters on the theory of pure tones, the nature of harmonics, the structure of instruments and the acoustics of the concert-room, was widely recognized as one of the most valuable introductions to the scientific aspect of music.

* See Memoir.

Music was indeed one of the strongest bonds of this second marriage and Jeans greatly enjoyed the concert tour in America in 1937 on which he accompanied his wife.

Three children were born of the marriage, Michael Antony (1936), Christopher Vincent (1939) and Katherine Anne (1944). With them Jeans renewed his youth, he even became a stamp-collector with his eldest boy and, like all fathers of young children in war-time, he found himself engaged in many kinds of unaccustomed domesticity.

In 1937 Jeans was awarded the Mukerjee Medal of the Indian Association and went to India to receive it. While he was there he took part in the meetings of the British Association and was awarded the Calcutta Medal of the Royal Asiatic Society of Bengal in the following year. In 1939 he received the supreme distinction of the Order of Merit and two years later his old college, Trinity, made him an honorary fellow.

In the early days of the war Cleveland Lodge was requisitioned by the military authorities and anti-tank trenches were dug across the grounds. Jeans and his family moved to Somerset and lived, first at Shepton Mallet and later near Wells, for the period of the war. Jeans disliked being cut off from London but he had his old friend Sir Richard Paget near him and he came up to the Royal Institution to lecture from time to time.

In the later part of the war Jeans was not a fit man. He was ill for some months while he was in Somerset and in January 1945 he had an attack of coronary thrombosis. He made a good recovery, but was obliged to give up many of his activities. He was well enough to attend the Royal Astronomical Society Club dinner in April and to go to Montreux for a summer holiday in the summer of 1946, but after his return he had a second attack of thrombosis and died at Cleveland Lodge on 16 September 1946. He is buried in Mickleham churchyard.

CHAPTER VII

SCIENCE IN JEANS'S BOYHOOD

BEFORE passing to an account of Jeans's technical and scientific achievements, it may be of interest to sketch the background of science in the days of his boyhood, more especially as Jeans himself was to write, at the age of fifty-six, a volume entitled *The New Background of Science.* And, as Jeans devoted the last eighteen years of his life to popular exposition, the best way of doing this would appear to be to take a brief survey of the state of popular science in the last half of the nineteenth century.

Let us take the lectures and writings of two renowned expositors of physics, Helmholtz in Germany and Tyndall in England.

The *Popular Lectures on Scientific Subjects* of Hermann von Helmholtz were published in an excellent English translation in two volumes in 1893. They consisted of addresses on sundry formal occasions, delivered to educated but not specialist audiences, and covered ground to which Helmholtz himself had made notable contributions. One group of addresses was concerned with the first law of thermodynamics (as it is now called), namely, the law of conservation of energy or, as Helmholtz termed it, the law of conservation of force. This great generalization appealed strongly to Helmholtz. It was the subject of his Carlsruhe address of 1862, 'On the Conservation of Force'; he had emphasized it in his Königsberg address of 1854 'On the Interaction of Natural Forces', and he was to dwell on it again in his Innsbruck address of 1869, 'On the Aim and Progress of Physical Science'. The possibility that the law of conservation of energy applied to all forms of energy had been outlined by Julius Robert Mayer in 1842; but it was the experiments of James Prescott Joule, published in 1843, which first established the strict equivalence of heat and mechanical energy. Joule's classical paper on this subject was dated 1849. It is evident from Helmholtz's insistence on the importance of this law that its power and generality were not fully realized by the audiences which he was addressing: It was necessary for him to pile example upon example.

More to the point in an account of Jeans's work is the fact that Helmholtz scarcely mentions the *second* law of thermodynamics in these addresses. He is at pains to stress the impossibility of a perpetual motion machine of what we now call the *first* kind—one which generates useful work indefinitely from nothing; but though he devotes a short paragraph to the mention of Sadi Carnot and his results on the relation of useful work to heat transferred, Helmholtz nowhere in these lectures stresses the impossibility of a perpetual motion machine of the *second* kind—one which would utilize as mechanical work the equivalent of the total heat lost to a source of heat. We may conclude that a knowledge of the consequences of the second law of thermodynamics was by no means a part of the general popular scientific background at the time of Helmholtz's addresses. It is inconceivable that Helmholtz himself had not recognized the importance of the second law. But he evidently deemed it beyond expression in popular lectures, though Clausius had given expression to his famous aphorism in 1850. It was to remain for Jeans to preach the doctrine that the final state of the universe would be a dull, uniform level of lifelessness, a state of uniform temperature, what is called a 'heat-death', a state of physical 'democracy' in which all inequalities were swept away so that nothing ever happened.

So much for the background of thermodynamics. Let us turn to Helmholtz's exposition of another subject which Jeans also expounded, though he did not make original contributions to it. I refer to non-Euclidean geometry. Helmholtz's Heidelberg address of 1870 was entitled 'On the Origin and Significance of Geometrical Axioms'. In this address he explains with great felicity of language the differences between Euclidean or 'flat' space, spherical space (the invention of Riemann) and pseudo-spherical space or, as we should now call it, hyperbolic space (the invention of Bolyai and Lobatchewsky). Helmholtz describes what sensible impressions we should receive if we were located in these various spaces, and shows how we can develop from the concepts of spherical or pseudo-spherical worlds the actual look of such a world in different directions. The whole climate of thought in this address is modern, and apposite to the science of today, when we have seriously to reckon with the possibility that the space of the universe is, from one point of view, actually hyperbolic. Helmholtz even considers the laws of motion in such a world. His own object was to controvert the view of Kant that the properties of space are given to us by intuition; Helmholtz opposed to this his view that our acquaintance with space was wholly empirical, not a transcendental form given *a priori* before experience.

Helmholtz himself had done original work on the axioms of geometry. But it was the impression made on him by the fundamental investigations of Gauss and Riemann which was at the bottom of his interest in the matter. His conclusions were in line with later views of Poincaré, namely, that the axioms of geometry, including the axioms of non-Euclidean geometries, are compatible with any empirical geometrical content whatever, but that, as soon as a framework has to be found for the principles of mechanics, we obtain a system of propositions which has real import, capable of verification or disproof by empirical observation. Lastly, 'if such a system was to be taken as a transcendental form of intuition, there must be assumed a pre-established harmony between form and reality'. This foreshadows Jeans's conclusion, in his philosophical writings, that the great Architect of the Universe was a mathematician.

It is highly remarkable that an investigator of the calibre of Helmholtz, with his feet so firmly planted on the ground of empirical observation and induction from experiment, should have ventured to expound the subleties of 'curved space' to a non-technical audience, so long before the theory of relativity had been established. We are at least clear that the curvature of space was part of the background of science in the late nineteenth century.

But Helmholtz had powerful interests in other branches of science, and amongst these was one very near to Jeans's heart, namely, the origin of the solar system. Helmholtz devoted to this topic his addresses of 1871 at Cologne and Heidelberg, 'On the Origin of the Planetary System', in which he expounded in a qualitative way the nebular hypotheses of Kant and Laplace. He prefaced his lecture by a defence of a branch of study which has not only appealed to most races, but has also compelled them to construct their own cosmologies, saying that the question of the beginning of things is closely connected with the perhaps more practical problem of the end of things: what may be formed, may also pass away. He pointed out further that Immanuel Kant was by instinct and inclination primarily a *natural* philosopher, the writings of his early period (to his fortieth year) belonging mostly to genuine natural philosophy, and that he made philosophy his later interest owing to the tone of thought prevalent at the time and for want of means to carry out independent scientific research. Kant's *General Natural Philosophy and Theory of the Heavens* appeared when he was thirty-one, whilst his *Critique of Pure Reason* appeared when he was fifty-seven. Thus, Helmholtz was disposed to allow more to Kant, the man of science, than Jeans was afterwards disposed to allow. Helmholtz at the same time gave independent credit to Simon, Marquis de Laplace, the mathematician who, after his magnificent contributions to celestial mechanics and to the problem of the stability of the solar system, allowed himself to speculate on its origin.* It was not, however, random speculation, but a theory put forward to explain five remarkable regularities in the solar system, namely: (1) the movements of the planets in the same sense and almost in the same plane; (2) the movements of the satellites in the same sense, with few exceptions; (3) the movements of rotation of the planets and the sun in the same sense, and in the same sense as their orbits; (4) the small eccentricities of the orbits of the planets and satellites; and lastly, (5) the large eccentricities of the cometary orbits. Helmholtz took the view that new facts accruing since the days of Laplace have added material confirmation to the nebular hypothesis. He also argued that the testing of the nebular hypothesis would be evidence as regards any limits to the validity of the laws of nature—the question whether they have always held in the past, and whether they will always hold in the future. It should tell whether our conclusions from circumstances as to the past, and as to the future, imperatively lead to what Helmholtz considered an impossible state of things, that is, to a beginning which could not have been due to processes known to us, in short, to a *creation*.

This is all in the most modern spirit, and reads just as freshly today as when it was first uttered (1871). Primacy of place he gave to the law of gravitation, bringing all bodies of the universe into connexion with one another, and culminating (a century ago) in the prediction of the existence of the planet Neptune at the hands of Leverrier and Adams. He pointed also to the prediction of the existence of a companion to Sirius, then fairly recently confirmed by Alvan Clarke. He pointed to the presence of asteroids and meteorites within the region of the solar system, as vestigial traces of the original nebula. He pointed to the possibility of the presence of obscuring matter within the confines of the original solar system, and he attributed the source of solar energy to the conversion of gravitational potential

* Laplace, *Exposition du Système du Monde*, 1835, p. 464.

energy into heat energy. And so he was led to his famous estimate of the age of the sun and its future continuance: 'the heat which the sun could have previously developed by its condensation would have been sufficient to cover its present expenditure for not less than 22,000,000 years of the past'; and 'the sun will probably still continue in its condensation, and this would develop fresh quantities of heat, which would be sufficient for an additional 17,000,000 years at the same intensity of sunshine as that which is now the source of all terrestrial life.'

By these various paths, Helmholtz was led to the same primitive conditions. He quoted confirmatory evidence, from the observed bright-line spectrum of the gaseous (galactic) nebulae, that the primitive state of all matter might have been gaseous. And he referred to the so-called planetary nebulae with the central stars, as evidence that similar processes might be going on in the heavens today. He instanced ex-novae, as being probably dead suns. This led him to evolution—Darwinism was still almost a novelty. He contemplated indeed in this lecture the possibility that all living creation on this earth might cease as the sun cooled (a 'first law' effect, not a 'second law' effect), but rejoiced to think that in spite of this the individual might still continue to struggle, to perfect his intelligence, to bear without fear the thought of the possibility that the thread of his own consciousness might one day break. He compared the individual, with his changing body (Helmholtz was primarily a physiologist), to a flame, or to a wave, whose constituents are renewed continually. Thus, by the writings of Helmholtz as by those of our own Kelvin, one of the main problems of cosmogony, the origin of the solar system, about which Jeans was to write so eloquently, was well before the scientific public in the latter half of the nineteenth century.

Lastly, in his physiological offices, Helmholtz was much concerned with the nature of sense-perception, with the empirical association of data of sense with the nature of the objects perceived and the general ideas of cognition. Like many of the greatest natural philosophers, Helmholtz was driven to study the bases of epistemology, its relation to sound and sight, and to the structure of the ear and eye. These were also to interest Jeans in his later writings.

We see that Helmholtz in his popular lectures developed his ideas on four topics that were to be Jeans's life-interests: (1) the laws of thermodynamics (though Helmholtz stressed only the first law, the law of conservation of energy); (2) the laws or axioms of geometry and the possibility of non-Euclidean geometries; (3) the origin and evolution of the planetary system; (4) the nature of cognition.

I have no knowledge whether Jeans read Helmholtz's lectures, and hence whether they influenced him directly. But they show that the late nineteenth-century background of general scientific knowledge was rich soil for a plant like the youthful Jeans. In the domains we have considered, there was not a great difference between the nineteenth- and twentieth-century backgrounds.

But let us turn to an English expositor, John Tyndall, as he appears in his *Fragments of Science* (2 vols., eighth edition, 1892). Here we are in a much more recognizably nineteenth-century background. Titles taken at random from the first volume of the essays or discourses include 'The Constitution of Matter' (*Fortnightly Review*, 1863), 'Radiation' (Rede Lecture, Cambridge, 1865) (a Lecture which Jeans in his turn was to give), 'On Radiant Heat in Relation to the Colour and Chemical Constitution of Bodies' (Royal Institution discourse, 1866), 'The Sky' (*Forum*, 1868), 'Contributions to Molecular Physics' (supplement to the Rede Lecture, Royal Institution, 1864, though of earlier date), 'The Physical Basis of Solar Chemistry' (Royal Institution discourse, 1861), 'On Force' (Royal Institution discourse, 1862). Tyndall, like Helmholtz, was preoccupied with the first rather than the second law of thermodynamics, and used the word 'force' where we should say energy. But his discourse on radiation was entirely concerned with the macroscopic aspects of radiation, not with the supreme theoretical question of the partition of the energy of equilibrium radiation in wave-length. Radiation was indeed to Tyndall the 'communication of vibratory motion to the ether', but he had no inkling that the corpuscular theory of radiation, favoured by Newton, was to be revived by Einstein. He was chiefly concerned in investigating whether the state of chemical combination of atoms influenced their reaction to radiation. But here too his experiments were macroscopic in conception.

Tyndall's *Heat, a Mode of Motion,* of which the sixth edition is prefaced by a note dated 1880, is in the form of seventeen lectures, devoted to the science of heat and especially the conservation of energy. But it is characteristic of the change of emphasis, which the twentieth century was to bring, that in this work the only reference to the second law of thermodynamics is a note attached to the 1880 preface, occasioned by the republication in 1878 of Sadi Carnot's celebrated *Reflexion sur la Puissance Motrice du Feu,* first published in 1824. But even so, Tyndall appears not to grasp the point of Carnot's investigation, namely that it contains the statement of a new law of thermodynamics. Tyndall was more concerned to defend Carnot from the accusation of falling into error by his initial adoption of the *caloric* theory of heat as a substance; on the evidence of new material contained in an appendix to Carnot's original memoir when it was republished in 1878,* Tyndall remarked that Carnot, before his death in 1830, had clearly disengaged his mind from this assumption. This to Tyndall was the important point, though it is not the aspect of Carnot's memoir to which the twentieth-century physicists, typified by Jeans, attach importance. Thus

* My own reprint of Carnot's memoir, dated 1912, does not contain the appendix referred to by Tyndall.

Tyndall, like Helmholtz, saw thermodynamical principles as substantially confined to the consequences of the first law of thermodynamics only. In a volume on heat of 572 pages the word *entropy* does not occur.

Likewise in his *Six Lectures on Light* (fourth edition, 1885), which is the collection in book form of lectures delivered in the United States in 1872–3, Tyndall is silent on the concept of 'black' or equilibrium radiation, though he does emphatic justice to the discoveries of Kirchhoff and his explanation of Fraunhofer absorption lines of 1862–3. On the other hand Tyndall defends with almost dogmatic emphasis the hypothesis of the ether,* when he says that the scientific imagination demands as the origin and cause of ether waves a particle of vibrating matter quite as definite (though it may be recessively minute) as that which gives origin to a musical sound. 'Such a particle we name an atom or a molecule.'

There were many other characteristics on the side of biological science to which I have scarcely referred. There were the lectures of Huxley, and there was Tyndall's great Belfast address to the British Association of 1874 in which he makes an eloquent and impassioned defence of the Darwinian theory. In this Belfast address he traces the whole philosophy of science, to the stage it had then reached, from Democritus through Lucretius, Bishop Butler and Kant to Clerk Maxwell; one may refer especially to his imaginary dialogue between Lucretius and Butler. Each generation of men of science has to relearn its history of science from the lips and pen of one of its contemporaries, and Jeans was to say over again, in twentieth-century idiom, many of the things said by Tyndall in his Belfast address.

* See in particular the discourse to the British Association at Liverpool of 1870. 'The Scientific Use of the Imagination'.

CHAPTER VIII

THE PARTITION OF ENERGY

It was shown in 1861 by G. Kirchhoff* that in an enclosure at temperature T, the state of the field of radiation depends only on this temperature T, and does not depend on the optical properties of the substances that happen to be present in the enclosure. This state of radiation is called complete or equilibrium radiation, or black-body radiation. It was one of the primary objects of theoretical physics in the nineteenth century to determine this characteristic state of radiation by calculation.

In the preceding paragraph I have stated the broad facts, so that the reader may see the issue. To give these facts their quantitative form, certain refinements of statement are needed. What Kirchhoff actually showed, by means of thermodynamic arguments, is as follows. Let the enclosure contain substances capable of emitting and absorbing radiation of energy frequency ν. At any point P in the enclosure, let the specific intensity of radiation for frequency ν be I_ν; that is to say, in a short time dt through an element of area dS containing P, in a cone of directions of solid angle $d\omega$ making an angle θ with the normal to dS, the flow of energy is taken to be $I_\nu d\nu dt dS \cos\theta d\omega$, $d\nu$ being a small range of frequencies surrounding ν. Further, let k_ν be the absorption coefficient of the material at P, j_ν the emission coefficient of the same material. These statements mean that a beam of radiation of intensity I_ν traversing a thin layer of the material of thickness dl is weakened by the amount $dI_\nu = -k_\nu \rho I_\nu dl$, where ρ is the density; and that the emission of radiant energy from a small element ρdv of volume dv in

* 'Untersuchungen über das Sormenspectrum und die Spectren der chemischen Elemente.' *Abhandlungen der königlichen Akademie der Wissenschaften zu Berlin* (1861); specially reprinted, with an enlarged appendix, 1862; Part 2, 1863.

time dt in directions included in $d\omega$ is $j_\nu \rho \, d\nu \, dt \, d\omega$. Then by ideal experiments with apertures, screens and mirrors inside the enclosure, Kirchhoff showed in effect (1) that the radiation is isotropic everywhere, that is, its intensity is independent of direction; (2) if μ_ν is the refractive index of the material at P, then I_ν/μ_ν^2 is the same everywhere within the enclosure; (3) that at any point $I_\nu = j_\nu/k_\nu$; (4) the value of I_ν/μ_ν^2 or $j_\nu/k_\nu\mu_\nu^2$ is the same for any two enclosures at the same temperature, independent of what substances they may contain, and so is a universal function of temperature and frequency.

We may in thought confine our attention to vacuous enclosures containing small particles of different substances. The refractive index μ_ν can then be taken as unity in the space between the particles. Then Kirchhoff's laws amount to saying that the ratio of the emissive power to the absorptive power of any material in thermal equilibrium at temperature T, in frequency ν, is the same for all materials and is a function of T and ν only.

For the benefit of the reader who finds the above mathematical statements tedious or difficult, it may be explained that the substances in an enclosure at a given temperature, surrounded by totally reflecting walls, are all exchanging energy with one another by means of waves of radiation. This radiation is a mixture of various frequencies, or wave-lengths, shorter wave-lengths or higher frequencies being more predominant at the higher temperatures. Each small element of material is maintained at a constant temperature by dint of absorbing, in any small element of time, as much radiant energy as it emits. In order to achieve this state of equilibrium, the environmental field of radiation must attain a certain intensity, neither too small nor too large: if the surrounding radiation had too small an intensity, emission of energy by an element of material would exceed absorption, the element would on balance lose energy, and its temperature would fall; if the surrounding radiation had too large an intensity, absorption of energy by an element of material would exceed emission, the element would on balance gain energy, and its temperature would rise. This argument by itself only shows that the *integrated* intensity of radiation at temperature T, i.e. the total intensity of all wave-lengths, must be definite. More refined thermodynamic arguments, depending on ideal experiments made with selectively reflecting and transmitting screens, were shown by Kirchhoff to lead to the result that there must be a balance of absorbed and emitted energy for each frequency separately. Otherwise it would be possible, without doing external work, to set up differences of temperature inside the originally isothermal chamber. Alternatively, if the balance did not exist, it would be possible to use the heat energy in the enclosure to perform mechanical work, without reference to external bodies. Both these results would contradict the second law of thermodynamics, and are therefore ruled out. The conclusion is, as before, that, provided the enclosure contains a mechanism for exchanging energy of all wave-lengths, the intensity of radiation in an enclosure at temperature T, for any given wave-length λ, is a function of T and λ only, and is independent of the particular chamber used, or the material in it used for effecting interchanges of energy. The proviso that the chamber must contain a mechanism for interchanging energy between different wave-lengths is obviously necessary, for otherwise we could have, in the enclosure, permanent specimens of different kinds of monochromatic radiation. When energy can pass from one frequency to another, the permanent state of radiation characteristic of the temperature T is set up.

By more refined ideal experiments, making use of the pressure of radiation predicted by Maxwell's theory of electromagnetic waves, it has been shown that the total intensity of equilibrium radiation at temperature T is proportional to the fourth power of T. This result is called

Stefan's law. By still more refined arguments, it has been shown by Wien that the resolution of the intensity of complete radiation into its separate wave-lengths is of the form

$$F(\lambda T)\,\lambda^{-5}d\lambda,$$

or

$$F(\nu/T)\,\nu^{3}d\nu.$$

But thermodynamic arguments alone are insufficient to determine the form of F. This can be shown to be due to the impossibility of devising a physical mechanism for drawing off a specimen of pure monochromatic radiation of a single frequency ν from a chamber containing complete radiation, the reason being that the act of motion of a piston working in a cylinder used to withdraw the monochromatic specimen through a selectively transmitting screen introduces changes of frequency by the Doppler effects occurring on reflexion at the moving piston, and so destroys the monochromatism.

Thus the calculation of the distribution of radiant energy with wave-length in complete radiation must depend on a detailed treatment of the mode of interaction of radiation with matter. It was to this difficult problem that Jeans repeatedly addressed himself in the first decade of the century.

The intensity of complete radiation on the classical mechanics was calculated by Jeans by a variety of methods which are summarized in the two editions (1914 and 1924) of his *Report on Radiation and the Quantum Theory*, which was produced for the Physical Society. One method was to calculate the emission and absorption of radiation by a Hertzian oscillator consisting of a single electron acted on by the oscillating electric field present in a beam of radiation. The condition for a steady state then determined the intensity of the beam, in terms of the mean kinetic energy of the resonator. According to the kinetic theory of gases, the mean energy per degree of freedom is $\frac{1}{2}RT$, where T is the temperature and R the gas-constant. Inserting this in the result of the previous calculation, Jeans found for the distribution or partition of radiant energy the formula

$$\frac{RT}{\pi^2 C^2} p^2 dp,$$

where p is proportional to frequency (it is in fact equal to $2\pi C/\lambda$), c is the velocity of light and λ the wave-length. The formula then gives for the partition of radiant energy at temperature T

$$8\pi RT\lambda^{-4} d\lambda.$$

Another classical method used by Jeans was to calculate the radiation from *free* electrons. (In the previous case the electron was *bound*.) He determined the condition that the interaction of the electron with the field of radiation should leave the partition of energy unaltered. This was a more difficult calculation, and involved the solution of an integral equation. Jeans was able to solve this, and obtained the same partition law as before.

Yet a third method was to calculate the radiation on the classical theory from electronic orbits. He again reached the same result. Of course, if the calculations were conducted properly, this was bound to be so, in accordance with Kirchhoff's result that the intensity of radiation in equilibrium with matter in an enclosure is independent of the particular properties of the matter present. The agreements showed that no mistakes had been made.

Jeans then proceeded to give the physical interpretation of this partition law. He considered any finite volume of a homogeneous medium, and calculated the number of free vibrations (with wave-lengths lying between given limits) of which this block of material would be capable. From dimensional considerations, this number must be of the form

$$C\nu\lambda^{-4} d\lambda,$$

where C is some constant. For sound vibrations in a gas, it can be shown that $C = 4\pi$; for light vibrations in free ether, (assuming an ether to exist) $C = 8\pi$; for elastic solid vibrations, $C = 12\pi$. The reason that these three values are in the ratio 1 : 2 : 3 is clear. In a gas, only waves of compression and rarefaction are possible; in the ether, only transverse vibrations are possible, and of these there are two sets, corresponding to the two possible components of polarization; in an elastic solid, both longitudinal (compressional) and transverse waves are possible, making three types altogether. In the radiation formula

$$8\pi RT\lambda^{-4}d\lambda$$

for the energy per unit volume, the average energy per vibration must accordingly be RT. This is a particular case of the general theorem of the *equipartition of energy.*

A considerable section of Jeans's *Dynamical Theory of Gases* is devoted to establishing the general theory of the equipartition of energy according to classical mechanics. This theory may be stated thus. Consider any dynamical system defined by a large number n of generalized coordinates $q_1, q_2, \ldots, q_n$. Corresponding to these there will be a set of conjugated momenta, $p_1, \ldots, p_n$. Imagine the state of the system at any one instant depicted by plotting the point $(q_1, \ldots, q_n; p_1, \ldots, p_n)$ in $2n$-dimensional space. As the system follows out its natural motion, this point will move and describe some trajectory in the generalized space. The whole of the space may be supposed filled by a 'dust cloud' of representative points. There is a well-known theorem, called Liouville's theorem,* which states that the density of any group of representative points of the dust-cloud does not alter as we follow its motion along a trajectory. For example, if the moving representative points are initially distributed with uniform density throughout the space, they will remain of uniform density for ever.

Suppose now (said Jeans) it is found that after a steady

* This can be proved from the Hamiltonian equations of motion of the system.

state has been reached (in which the influence of initial conditions is obliterated) the system nearly always possesses some definite property P. This might *a priori* arise from either of two sets of circumstances: either the representative points tend to cluster—tend to be trapped, as it were—in those regions of the generalized $2n$-dimensional space for which the property P obtains; or the property P is common to nearly the whole of the generalized space. Liouville's theorem shows that the first alternative cannot hold. Hence the property P must hold good throughout nearly the whole of the generalized space. Thus properties to be looked for in the steady state of a system are those which hold good 'almost everywhere' in the generalized space.

Such a property is given by the following exposition. Consider the expression for the *energy* of the system when it has been reduced to a sum of squared terms. The number of such terms will be very large, and we can discuss the contribution they make to the energy in terms of their average sizes. If we examine a group of n_1 such terms, and another group of n_2 such terms, then it can be shown that the ratio of the sum of the n_1 terms to the sum of the n_2 terms is n_1/n_2. In other words, the average value of a term in the group of n_1 is equal to the average value of a term in the group of n_2. This is equivalent to saying that every squared term in the expression for the energy has the same average value. This average value can be shown by thermodynamic arguments to be proportional to the absolute temperature T of the system, and its value is taken to be $\frac{1}{2}RT$, the constant R being the same as that occurring in the statement of the equation of state of a perfect gas, $p/\rho = (R/\mu)\,T$, where μ is the mass of a molecule. This is the famous theorem of the equipartition of energy in classical mechanics—the result that on the average no one degree of freedom contains more energy than any other. It is true that it is only established for a finite, even if very large, number of degrees of freedom. The ether in a finite volume has a finite number of degrees of freedom between any given limits of wave-length, but an infinite number of degrees of freedom *in toto*. There is, however, no reason to suspect the applicability of the theorem to the ether by classical mechanics, since the result of so applying it has been obtained by so many independent methods.

The Rayleigh-Jeans formula, $8\pi RT\lambda^{-4}d\lambda$ for the radiant energy per unit volume between wave-lengths and $\lambda+d\lambda$, cannot, however, apply in nature for all wave-lengths however small, since it would imply that the whole of the energy was located in indefinitely small wave-lengths, in disagreement with experience. This is the *ultra-violet catastrophe*, already described in Chapter III, which nature avoids. It was to explain this avoidance that the quantum theory came into being.

Jeans went on to survey possible ways of avoiding the ultra-violet catastrophe without contradicting classical mechanics. He pointed out that the classical theory required that the degradation of energy into the shorter wave-lengths should occur before a steady state was reached. The possibility might therefore arise that the observed final state in an enclosure was not one of thermodynamical equilibrium. In that case, different mechanisms of interaction of matter with radiation need not all lead to the same law of partition of radiant energy. Jeans showed that in the case of an accelerated free electron, if its environment is almost devoid of radiant energy, the law of partition of energy emitted, in terms of wave-length λ, was found to be, instead of

$$8\pi RT\lambda^{-4}d\lambda,$$

of the form

$$8\pi RT\lambda^{-4}f(\rho/\lambda)\,d\lambda,$$

where $\rho/2\pi c$ is comparable with the duration of the time of collision between the electron and the atom accelerating it. Since the ultra-violet catastrophe would be a phenomenon of short wave-lengths if it occurred, we may assume that the

Rayleigh-Jeans formula itself holds good in the limit of very long waves; and Jeans showed that the function f would tend to zero like $e^{-e/\lambda}$ for λ small. A similar result was derived by J. J. Thomson. To reconcile this result with Wien's law, ρ must vary inversely as the absolute temperature T. This was shown to give rise to an unlikely law for the force between an electron and an atom. Nevertheless, the two formulae for the partition of radiant energy are compatible, for λ large and λ small respectively, with the observed partition law, which was given by Planck in the form

$$8\pi RT\lambda^{-4}d\lambda\frac{x}{e^x-1},$$

where $x=h\nu/RT=hc/\lambda RT$, and h is a constant new to physics. It is now known as Planck's constant.

The simplest interpretation of this law is that it represents a modification of the theorem of equipartition of the following kind. Instead of allowing the energy of a system to vary continuously, suppose that out of a great number M of vibrations, N have zero energy, $Ne^{-\epsilon/kT}$ have energy ϵ, $Ne^{-2\epsilon/kT}$ have energy 2ϵ, and so on. These are in the same ratio as the numbers that would be given by the equipartition theory, but intermediate amounts of energy are supposed to be inadmissible. Then the total number of vibrations M must be given by

$$M=N(1+e^{-\epsilon/kT}+e^{-2\epsilon/kT}+\dots)=\frac{N}{1-e^{-\epsilon/kT}}.$$

On the other hand the total amount of energy is

$$\begin{aligned}E&=N\epsilon(e^{-\epsilon/kT}+2e^{-2\epsilon/kT}+3e^{-3\epsilon/kT}+\dots)\\&=\frac{N\epsilon e^{-\epsilon/kT}}{(1-e^{-\epsilon/kT})^2}.\end{aligned}$$

Hence the average amount of energy per vibration is

$$\frac{E}{M}=\frac{\epsilon}{e^{\epsilon/kT}-1}.$$

If we take M to be the number calculated earlier,

$$M = 8\pi\lambda^{-4}d\lambda,$$

then $$E = 8\pi\lambda^{-4}d\lambda \frac{\epsilon}{e^{\epsilon/kT}-1}.$$

This can be reconciled with observation only by taking

$$\epsilon = h\nu,$$

where $\nu = c/\lambda$ is the frequency of a vibration. Thus the experimental validity of Planck's formula means that the hypothetical resonators in material systems can only exchange energy with the ether, in any given wave-length λ or frequency ν, discontinuously, in units of amount $h\nu$.

Such was the beginning of the quantum theory. We do not follow here its enormous development into a quantum theory of atomic structure at the hands of Bohr, of dynamics at the hands of Heisenberg and Dirac, or of statistical mechanics at the hands of Fermi-Dirac and Einstein-Bose, as Jeans took no part in such developments. But 'c'est le premier pas qui coûte', and it is difficult at the present stage of the evolution of physics to realize the reluctance of the physicists of the generation immediately preceding our own to admit that Newtonian mechanics was not ultimately valid in all circumstances. The reluctance in the case of relativity has not been entirely overcome even today.

CHAPTER IX

ROTATING FLUID MASSES

I PROPOSE in this chapter to sketch in as non-technical language as possible the classical subject of the forms of equilibrium of rotating, gravitating fluid masses and their stability, as it was when Jeans began to make contributions to it, and the nature of Jeans's contributions.

Suppose we consider a mass of incompressible liquid, spinning about an axis and isolated in space. What form will it assume, and how will the form change, if at all, as the mass shrinks? By the theorem of the conservation of angular momentum, as such a mass shrinks, and its moment of inertia consequently decreases, its angular velocity will increase. The problem can therefore be reduced to that of the forms of equilibrium of a rotating homogeneous mass as its angular velocity increases from zero upwards. When the angular velocity is zero, the form is evidently that of a sphere (though this is by no means as easy to prove as it looks). It was shown by Newton, and more particularly by Maclaurin, that as angular velocity sets in, the form is initially that of an ellipsoid of revolution, an oblate spheroid in fact, with the shorter (polar) axis lying along the axis of rotation. As the angular velocity increases, the polar axis shortens and the equatorial axes lengthen. This process goes on until the angular velocity ω reaches a certain maximum (given by* $\omega^2/2\pi G\rho = 0{\cdot}2247$, corresponding to an eccentricity of meridian section equal to 0·93). For higher values of the angular velocity, no forms of equilibrium of the type of a spheroid are possible. But a second set of spheroidal figures are possible, corresponding to decreasing angular velocity but still increasing eccentricity, until in the limit the form assumed is that of a flat disk of

* G is the gravitational constant, ρ the density.

very large radius and zero thickness, rotating very slowly about an axis through its centre normal to its plane.

Not all these configurations of relative equilibrium, however, are stable. The configurations for which $\omega^2/2\pi G\rho$ has diminished again beyond the value of 0·2247 are all unstable, in the ordinary sense of that word. Thus all the disk-like configurations are unstable. But some of the configurations on the ascending branch of angular velocity are unstable in another sense. This type of instability occurs when dissipative forces are present. It arises in the following way.

A purely statical system is stable when its potential energy in the given configuration is a minimum. The reason is that, in any displaced motion, the sum of the potential and kinetic energies must remain constant; and so, since the potential energy cannot decrease below a certain limit (namely the minimum value corresponding to the given configuration of equilibrium), the kinetic energy cannot increase above a certain amount, which for a small displacement remains small. Hence, the configuration is stable. When, however, the system is not a statical one, when, for example, as in the present context, we are considering displacements from a given state of steady rotation, there exists a similar function of position (involving the total moment of momentum) which has to be stationary in the given state of steady motion. In any disturbed motion, the sum of this function and a function analogous to the kinetic energy (quadratic in the relative velocities) must remain constant. But now the presence of viscous forces will cause, in the disturbed motion, a diminution in the total energy, by conversion of the energy of rotation into heat; and an indefinite increase in the relative kinetic energy may be compatible with constancy of the combined functions mentioned earlier, by secular decrease of the function analogous to the potential energy. The possibility, so to speak, of drawing on the kinetic energy of rotation to augment the kinetic energy of relative motion in the motion following a displacement, means that more drastic conditions are required to ensure stability against the secular conversion of kinetic energy of rotation into heat energy. This form of instability was called by Thomson and Tait 'secular instability'. Jeans remarks that it is clear that a system which is ordinarily (i.e. neglecting dissipative forces) stable may or may not be secularly stable, but a system which is ordinarily unstable is necessarily secularly unstable.

As an example of a case of secular instability in the presence of dissipative forces, Lamb gave the case of a particle moving on the interior of a spherical bowl. If the velocity of the particle is sufficiently high, a steady state of rotatory motion is possible in which the particle is not at the lowest point of the bowl. If the bowl is smooth, this configuration of steady motion is stable; but if the bowl is rough, that is, if dissipative forces are present, the particle will eventually descend to the lowest point, the kinetic energy of the motion of descent being drawn from the kinetic energy of the original rotatory motion. The eccentric position of the rotating particle is then secularly unstable.

Returning to the case of the rotating mass of fluid, it has been shown that the oblate spheroids become secularly unstable before the angular velocity attains its maximum. The value of $\omega^2/2\pi G\rho$ at which this occurs is 0·18712 (less than 0·2247), when the eccentricity (e) of the meridian section is $e = 0{\cdot}81267$. In actual practice, then, oblate spheroids do not correspond to stable rotating masses after $\omega^2/2\pi G\rho$ has reached the value 0·18712; and the question arises what happens to the rotating mass as its angular velocity increases beyond the value corresponding to $\omega^2/2\pi G\rho = 0{\cdot}18712$. The answer is that a new set of configurations of relative equilibrium make their appearance, in the form of ellipsoids, having three unequal axes. They are known as Jacobian ellipsoids, having been discovered by Jacobi in 1834. They join continuously on to the

Maclaurin spheroids, the limiting Maclaurin spheroids being the first Jacobian ellipsoid, the particular ellipsoid with the two equatorial axes equal in length ($a = b = 1{\cdot}7161\,c$). Such a point, where a new series of configurations of relative equilibrium diverges from another series is called a *point of bifurcation.* It was shown by Henri Poincaré in 1885 that at a point of bifurcation there may be an exchange of

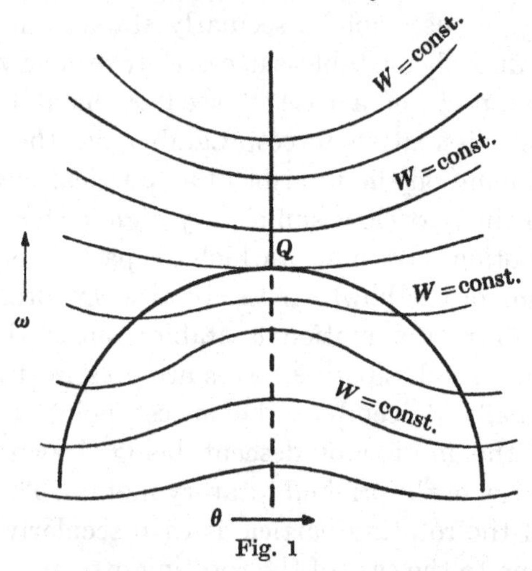

Fig. 1

stability. In the present case, the stability passes to the Jacobian ellipsoids.

The theory of exchange of stability at a point of bifurcation, which was discovered by Poincaré analytically, was made intuitively obvious, in the case of purely statical systems, by Jeans in his Adams Prize Essay, using a series of diagrams. In such statical systems, the configurations of equilibrium occur at places where the potential energy W is stationary. At such a point, if the surface, $W = \text{const.}$, is plotted against the generalized co-ordinates, the tangent plane to the surface, $W = \text{const.}$, is horizontal. If the position of this surface depends on a parameter (like the angular velocity ω, in the non-statical case) there will be a family of such surfaces for different values of ω, and each point where the tangent is horizontal (i.e. at each 'level point') will give a configuration of equilibrium. For continuously varying ω, the point of equilibrium will trace a locus which is called

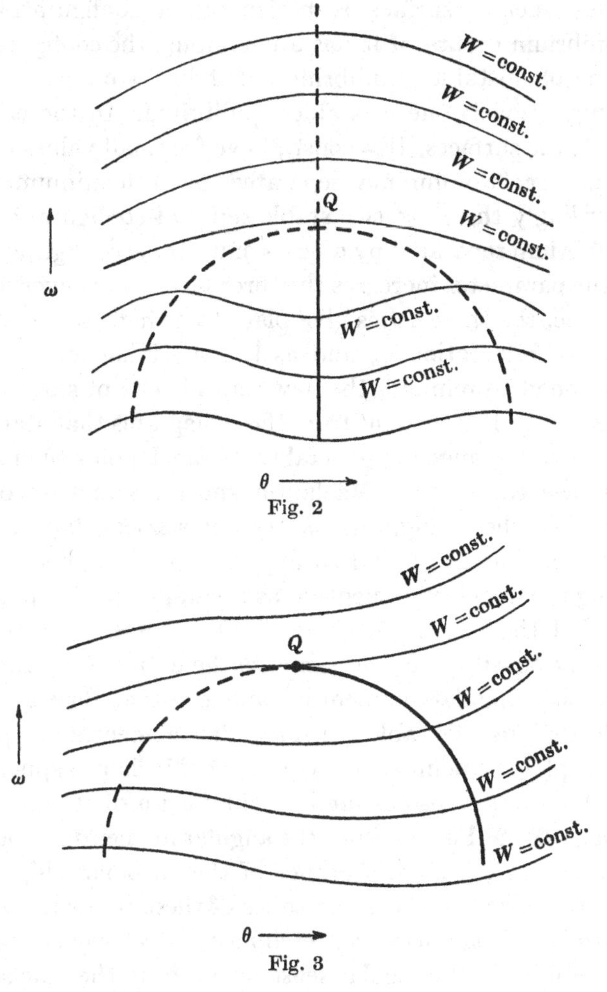

Fig. 2

Fig. 3

—— Stable configurations. — — — Unstable configurations.

a 'linear series' of equilibrium configurations. This situation is illustrated in Figs. 1, 2 and 3, which are taken from Jeans's Adams Prize Essay.

In these a typical generalized co-ordinate θ is taken as abscissa and a varying parameter defining the system is taken as ordinate, and the surfaces, W = const., are plotted. Where such a surface is horizontal, a configuration of equilibrium occurs; if it has a maximum, the configuration is one of unstable equilibrium; if it has a minimum, the configuration is one of stable equilibrium. In the case of Fig. 1, the surfaces, W = const., have for small values of the parameter two minima separated by a maximum, and accordingly there are two stable series of configurations of equilibrium separated by a series of unstable configurations. As the parameter increases the three level points eventually coincide, the three series give place to a single series at the point of bifurcation Q, and as the surviving level points correspond to minima, the new series is one of stable configurations. It is evident from these diagrams that stability will be interchanged in general at a branch point such as Q.

In the case of the Maclaurin spheroids and Jacobian ellipsoids, the configurations are not statical, but similar considerations are found to apply. Jeans supplied tables giving the angular momentum as a parameter. He pointed out that the case of constant angular momentum and increasing density was the same mathematically as that of increasing angular momentum and constant density. He could then use his tables with angular momentum as parameter, plot the points corresponding to Maclaurin spheroids and Jacobian ellipsoids, and he obtained a diagram of the form shown in Fig. 4. Since the angular momentum goes on increasing with the appearance of the Jacobian ellipsoids, the curve showing the linear series of these bends upwards, and so (as is easily seen by sketching in the relevant surfaces) the stability in the secular sense passes from the Maclaurin spheroids to the Jacobian ellipsoids.

It was shown by Poincaré that as we pursue the Jacobian series, another point of bifurcation is reached, at which still another series of equilibrium configurations make their appearance. A furrow develops surrounding the long axis of the Jacobian ellipsoids, not symmetrical with respect to the centre of the system; these are called the 'pear-shaped' figures of equilibrium. At this point of bifurcation, the Jacobian ellipsoids must lose their stability. The question is, does the stability pass to the pear-shaped configuration, or does it disappear altogether?

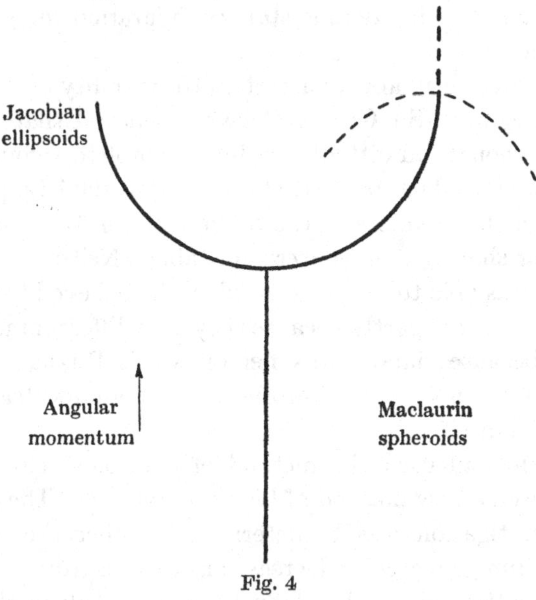

Fig. 4

The importance of this question in relation to the evolution of some of the heavenly bodies is evident. Double stars may have evolved from single masses by fission, but we should like to know whether the process of fission took place slowly or catastrophically. At one end of the series of possible figures of relative equilibrium we have the Jacobian ellipsoids. At the other end, we have the stable system consisting of two detached masses revolving around one another, and forming a stable system, though each will cause a tidal distortion in the other. Are these two ends of the process connected by figures of equilibrium, of which the earlier ones might be the pear-shaped figures? Or does the passage from the Jacobian ellipsoids to the configuration consisting of two discrete masses occur as a cataclysm? If the pear-shaped figures are unstable, that would indicate that at the end of the Jacobian series, a cataclysm should occur. If on the other hand the pear-shaped figures are stable, this would at any rate be some indication that a passage to the double-star configuration may occur continuously.

Poincaré was unable to ascertain the stability of the pear-shaped figures. Sir George Darwin* believed that he had shown, though admittedly without complete rigour, that the pear-shaped figures were stable. But in 1905, Liapounoff stated in the memoirs of the St Petersburg Academy that the pear-shaped figures were unstable. Neither of these writers was able to say exactly where he believed the other had gone wrong, partly because they used different methods, partly because Liapounoff's memoir was in Russian. To the removal of this unsatisfactory state of affairs Jeans addressed himself.

Darwin had used the method of ellipsoidal harmonics. Jeans used a new method of his own invention. The goal of the investigation was to determine whether the angular momentum increased or decreased as we pass from the point of bifurcation on the Jacobian linear series along the new series of pear-shaped figures. It will be seen from the diagrams above that if the curve of angular momentum decreases from the point of bifurcation R, along the pear-shaped series, this series must, from Poincaré's general principles, be unstable; but if it increases, the pear-shaped figures will be stable.

With this goal in view, Jeans's first care was to calculate the gravitational potential of a distorted ellipsoid. This is an essential step in the location of the point of bifurcation on the Jacobian series. The amount of the distortion was

* *Collected Papers* (1902, 1908, 1909), 3, 317.

specified by a certain function of position added in to the left-hand side of the ordinary equation of an ellipsoid, multiplied by a small parameter e. By some masterly if not perfectly rigorous analysis, Jeans succeeded in calculating, as a closed expression, the gravitational potential at the boundary of the distorted ellipsoid and at interim points. The advantage of Jeans's expression was that it was similar in form to, though more complicated than, the usual expression for the potential of an undistorted ellipsoid, to which it reduced for $e=0$. Jeans's results for the *position* of the point of bifurcation on the Jacobian series agreed with Darwin's, though it was obtained by a different method.

It is evident that since it is only in the vicinity of the point of bifurcation that it is necessary to ascertain the variation of angular momentum, attention can be confined to small values of the distortion-parameter e. It was found by Jeans that, if approximation were carried out only as far as terms in e^2, the analysis was not adequate to isolate a determinate linear series of pear-shaped figures of equilibrium. Instead, as far as terms in e^2 were concerned, the two series (the Jacobian ellipsoids and the pear-shaped figures) lost their identity near the point of bifurcation; instead of linear series in the diagram, the configurations of equilibrium occupied, or appeared to occupy, an area. Hence a higher approximation is required before one can isolate the series, let alone ascertain the stability or instability of the new series. Jeans remarked that this was the point at which Darwin was unfortunately misled. To quote Jeans:*

Sir George Darwin seems to have carried out his investigation under the impression that there would be a unique configuration of equilibrium when the calculations were carried out as far as e^2, and this led him to introduce a spurious condition of equilibrium, the effect of which was to limit him to one of the doubly infinite series we have discovered. In point of fact, Darwin's extra condition of equilibrium could only be satisfied by assigning

* *Problems of Cosmogony and Stellar Dynamics* (1919), p. 92.

to a special value,* namely $\zeta = -0{\cdot}015988\,e^2$, and this value gives a figure whose angular momentum is greater than that of the undistorted ellipsoid. Darwin accordingly announced the pear-shaped figure to be stable.

But we shall now see that this special value for ζ makes it impossible to carry the linear series on to third order terms at all. The condition that it shall be possible to carry on the series to third order terms requires that ζ shall have a special value, but this special value is not the one assumed by Darwin; it is a value which shows the pear-shaped figure to be unstable, as we shall now see.

Jeans proceeded to calculate the pear-shaped series as far as the third-order terms. He accordingly added to the left-hand side of the ordinary equation of an ellipsoid terms in e, e^2 and e^3, each governing an appropriate polynomial in the co-ordinates, and then applied his previous work to calculate the gravitational potential of the new figure. This work is extremely formidable, and it is impossible here either to indicate its nature or to do justice to its complexity.† Suffice it to say that Jeans found that the equations determining the third harmonic displacement possessed a solution only when $\delta\omega^2$ took the particular value

$$\frac{\delta\omega^2}{2\pi G\rho} = 0{\cdot}0074231,$$

so that as we pass along the pear-shaped series, we have initially

$$\omega^2 = \omega_e^2 + 0{\cdot}0074231\; 2\pi G\rho e^2,$$

where ω_e^2 refers to the angular velocity of the critical Jacobian ellipsoid; and so ω^2 increases initially. The value of $\omega_e^2/2\pi G\rho$ being $0{\cdot}1419990$, as found by Darwin, we have now for the pear-shaped figure

$$\frac{\omega^2}{2\pi G\rho} = 0{\cdot}14200\;(1 + 0{\cdot}05227e^2).$$

* In the equation $\zeta = \dfrac{\delta\omega^2}{2\pi G\rho}e^2$, $\omega^2 + e^2\delta\omega^2$ is the value of ω^2 for the new configuration.

† The reader is referred to *Problems of Cosmogony and Stellar Dynamics*, pp. 93–9.

Jeans next calculated the square of the radius of gyration of the pear-shaped figure, in the form

$$k^2 = 0{\cdot}8441(1 - 0{\cdot}0937\,e^2),$$

whence it follows that the angular momentum M of the pear-shaped figure is connected with the angular momentum M_0 of the Jacobian ellipsoid by the relation

$$M = M_0(1 - 0{\cdot}06765e^2).$$

Hence M is less than M_0, the angular momentum decreases along the pear-shaped series, and the pear-shaped figures are unstable.

The instability of the pear-shaped series, thus demonstrated by Jeans, shows that a rotating mass cannot evolve by slow secular changes through a series of pear-shaped figures. Jeans remarked, as a sort of anti-climax, that this somewhat diminished the interest of the pear-shaped figures in the problem of cosmogony. Nevertheless, as he pointed out, it still remains of importance to obtain clear ideas about the nature of this series, since the series is a guide to the course of the cataclysm that must result when the Jacobian series are past.

Instead, however, of attempting the difficult problem of further study of the pear-shaped figures, Jeans sought earlier to throw light on the problem by examining the analogous but simpler problem in two dimensions, in which the three-dimensional bodies are replaced by cylinders. He wrote an important memoir on this subject.* The substance of this is reproduced in abbreviated form in the Adams Prize Essay, and Jeans's conclusion from the investigation runs:

> Thus we may assert with fair confidence that the two-dimensional series ends by fission into two detached masses, and in view of the close parallelism which we have discovered between the two-dimensional and the three-dimensional problem,

* *Philosophical Transactions of the Royal Society*, A (1902), **200**, 67–104.

it seems highly probable that the three-dimensional series also will end by a similar fission into detached masses.

Thus far we have followed Jeans's discussion of rotating gravitating masses alone in space. But in his Adams Prize Essay Jeans gave an exposition of other cosmogonic problems involving incompressible masses acted on by their own gravitation, and linked together the works of his predecessors. He classified these problems under three heads:

(*a*) the purely rotational problem (which we have already discussed);

(*b*) the tidal problem, in which a primary mass of incompressible fluid is acted on tidally by a secondary mass, assumed to be concentrated at a point at a finite distance. The problem is the shape and stability of the primary as the mass and distance of the secondary are altered;

(*c*) the double-star problem, in which two bodies are rotating round one another without any change of relative position. This is in effect a combination of the rotational and tidal problems.

The solutions of these problems were largely obtained by Jeans's predecessors, amongst the moderns especially Roche and Sir George Darwin. But Jeans's re-derivation of their results is a work of great beauty. The relationships of the various solutions are shown diagrammatically in Fig. 5.*

In this diagram each ellipsoidal configuration is represented by a point, whose co-ordinates are the two semi-axes a and b perpendicular to the axis of rotation (when there is an axis of rotation). The volume of the incompressible mass remaining constant, we have $abc = r_0^3$, where r_0 is the radius of the sphere of equal volume. This, the spherical configuration, is represented by the point S, cor-

* Reproduced, with additional information, from *Problems of Cosmogony and Stellar Dynamics*, pp, 50 and 86.

responding to zero angular velocity ω. For the Maclaurin spheroids, $a=b$, and consequently they are represented by the continuation $SBM'M$ of the line OS bisecting the angle between the co-ordinate axes. Through S there passes a

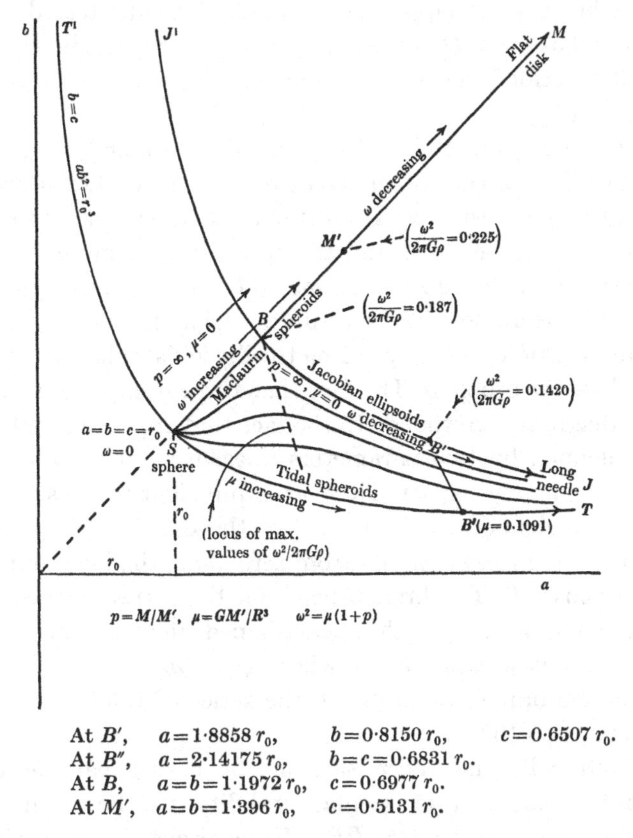

At B', $a=1{\cdot}8858\,r_0$, $b=0{\cdot}8150\,r_0$, $c=0{\cdot}6507\,r_0$.
At B'', $a=2{\cdot}14175\,r_0$, $b=c=0{\cdot}6831\,r_0$.
At B, $a=b=1{\cdot}1972\,r_0$, $c=0{\cdot}6977\,r_0$.
At M', $a=b=1{\cdot}396\,r_0$, $c=0{\cdot}5131\,r_0$.

Fig. 5. Configurations of relative equilibrium. (After Jeans, *Problems of Cosmogony*, fig. 7, p. 50). (Qualitative only.)

pseudo-hyperbolic locus $T'ST$ representing another set of spheroids—the prolate tidal spheroids for which the axes b, c in directions perpendicular to the direction of the secondary are equal. The hyperbola thus has for its equation $ab^2=r_0{}^3$. The point B is the point of bifurcation on the series of Maclaurin spheroids, and through this passes the locus $J'BJ$, giving the Jacobian ellipsoids, of three unequal axes.

The point infinitely distant along SM corresponds to the limiting Maclaurin configuration of a very thin slowly rotating disk of large radius. The point infinitely distant along the curve BJ corresponds to a very long needle-shaped configuration, the limiting Jacobian ellipsoid, also slowly rotating.

Along SB and as far as a point M', the angular velocity of the Maclaurin spheroids increases; it reaches a value corresponding to $\omega^2/2\pi G\rho = 0{\cdot}187$ at B, where the Jacobian ellipsoids branch off; and it attains a maximum at M', where $\omega^2/2\pi G\rho = 0{\cdot}224$. Along BJ the angular velocity steadily decreases. The curvilinear triangle $TSBJ$ bounds a number of loci corresponding to ellipsoidal solutions of the double-star problem. The fact that these occupy an area in the diagram corresponds to the fact that the configurations are defined by two parameters, measuring the mass ratio and the distance apart. These are equivalent to the symbols p and μ defined in the legend to the diagram. The parameter μ increases steadily from zero along the locus of tidal spheroids ST. The branch-point on the series of Jacobian ellipsoids at which the pear-shaped figures come into existence is specified by B, where $\omega^2/2\pi G\rho = 0{\cdot}142$. There is a similar branch point B'' on the series of tidal spheroids, where $\mu = 0{\cdot}1091$.

From what has been said in the earlier part of this chapter, the loci of secular stability in the rotational problem are the arcs SB, BB'. Jeans discussed the stability of the tidal and double-star configurations, but it would take us too far into technicalities to follow him here. It must suffice to say that only the earlier parts of the tidal series ST and the various double-star series are stable.

Jeans followed his account of configurations of equilibrium and their stability with an investigation of the type of cataclysmic motion that would occur on passage through a point of instability. He summarized his conclusions as follows:*

In all three problems [the tidal, rotational and double-star problem] we have found that the motion will consist of two parts. The first may be described as 'statical' or 'secular'; the second may be described as 'dynamical' or 'cataclysmic'.

In the Rotational Problem and in the Double-star Problem, there is a quite precise demarcation between the two types of motion. In the Tidal Problem, the two motions may gradually merge into one another....

In all three problems, the statical motion has been found to consist of a slow secular change of shape in which the body under consideration remains always of a spheroidal or ellipsoidal shape, except that in the tidal and double star problems (in which two masses are involved) the spheroidal or ellipsoidal shape of the primary may be slightly distorted by tides of third and higher orders raised by the secondary mass. In the Tidal Problem, the motion is through a series of prolate spheroids; in the Rotational Problem the motion is first through a series of oblate spheroids (Maclaurin's spheroids), and then through a series of ellipsoids (Jacobi's ellipsoids); in the Double-star Problem the motion is through a series of ellipsoids.

In all three problems, dynamical motion supervenes when the prolate spheroid or ellipsoid reaches a certain elongation. The motion results in the formation of a furrow or system of furrows on the elongated mass. In the Tidal Problem the furrowing process does not commence immediately, and there may be any number of furrows formed. In the two other problems, the furrows start to form at once and only one furrow is formed.

The final result of the dynamical motion appears in every case to be fission into detached masses, although a rigorous mathematical proof of this has not been obtained. In the Tidal Problem any finite number of detached masses may result; in the Rotational Problem the mass appears to divide into two bodies of unequal size; in the Double-star Problem the mass breaks up into a very great number of small masses.

Hence it appears highly probable that tidal action may produce

* *Problems of Cosmogony and Stellar Dynamics*, p. 137.

systems such as are seen in our own solar system and in the systems of Jupiter, Saturn, etc.; that increasing rotation may produce systems such as are seen in ordinary binary stars; and that the close approach of two stars revolving about one another may produce systems such as Saturn's rings and possibly the asteroids also.

Thus far Jeans had dealt with homogeneous incompressible masses only. But I have thought it worth while to give an account of this work in some detail because in a sense Jeans has written the final chapter on the subject. It is not that, cosmogonically, incompressible fluid masses are of primary importance; it is rather that the theoretical behaviour of incompressible masses has been a standing challenge to mathematicians, in the sense that, until their behaviour was fully known, it would be impossible to say what particular results would be the consequence of compressibility. They were required as a standard or norm of behaviour. And it is likely that Jeans's contributions to this field will rank amongst his more permanent achievements, come what may in the future development of cosmogony.

In his investigations of the forms of compressible masses, Jeans considered three types:

(1) Roche's model, consisting of a point nucleus of very great density, surrounded by an atmosphere of negligible density;

(2) A generalized Roche's model, with a finite incompressible nucleus;

(3) An adiabatic model consisting of a mass of gas in adiabatic equilibrium.

He pointed out that these could be related to one another and to the incompressible case by means of the following diagram (Fig. 6).

He then worked out the mode of break-up of the different types of system by rotation and by tidal action respectively. Configurations of type *A* break up by fission, as we have seen. Configurations of the type of Roche's model *B* break up, under increasing rotation, quite differently. They gradually become lens-shaped, with a sharp lenticular edge along the equator, and any further increase of rotation results in matter being thrown off from the equator in a continuous stream, owing to 'centrifugal force' exceeding gravity.

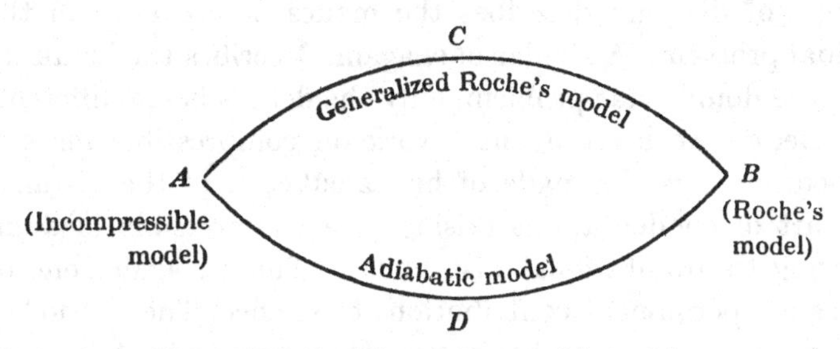

Fig. 6. Relationships of compressible models.

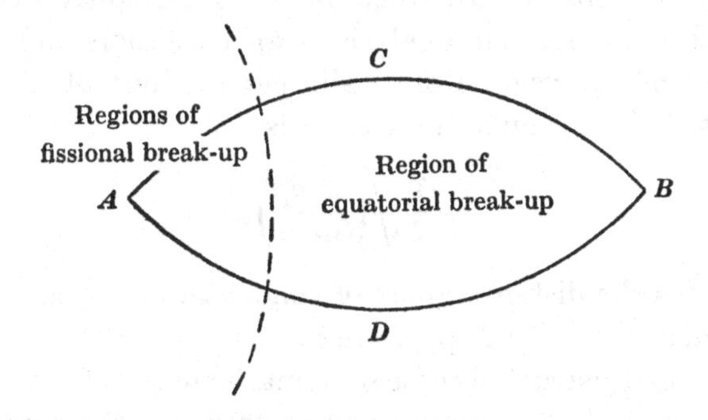

Fig. 7. Domains of different modes of rotational break-up.

Hence as we pass along the chain of models from *A* to *B*, whether via *C* or via *D*, there must come a point on each chain at which fissional break-up gives place to equatorial break-up. Jeans's investigations determined the critical point on each chain. It appeared that for all gases occurring in nature, the type of break-up would be equatorial; only compressible masses with an impossibly high ratio of specific heats could break up by fission. Jeans was therefore led to the diagram for rotational break-up (Fig. 7).

In the case of the tidal problem, there is a similar division into regions of fissional break-up and equatorial break-up, but the ejection of material in the latter case, instead of occurring regularly at all points on the equator, is confined to one or perhaps two 'conical points', in the direction of the tide-raising forces. Thus, much the same type of diagram describes the modes of break-up in the tidal problem. A similar one, again, describes the break-up in the double-star problem, only the details being different.

Before we leave Jeans's work on compressible masses, mention must be made of his calculation of the distance apart of condensations arising in a compressible medium owing to gravitational instability.* This is, again, one of Jeans's permanent contributions to science. The method of derivation is extremely elegant. He mentions in *Astronomy and Cosmogony* (p. 337, where he gives a slightly different derivation), that the analysis is an extension and modification of analysis he originally gave in one of his early papers.† The formula in question is

$$l=\frac{1}{2}\sqrt{\left(\frac{\pi}{G\rho}\frac{dp}{d\rho}\right)},$$

where l is the distance apart of condensations in an arm of a nebula of density ρ, pressure p.

Having constructed these solid mathematical foundations Jeans proceeded to apply them to elucidate the courses of evolution of different types of heavenly bodies. In an introductory chapter he had drawn attention to five main uniformities of structure in the universe at large. These were the spiral nebulae, the 'planetary' or ring nebulae, star-clusters, double and multiple stars and the planetary system. By as detailed a discussion as was possible in the actual state of knowledge at the time, Jeans showed that it was likely that the spiral nebulae had been formed by a

* *Problems of Cosmogony and Stellar Dynamics*, pp. 157–60.

† *Phil. Trans.* (1902), 199 A, 49, 'The stability of a spiral nebula'.

mode of equatorial break-up (due to rotation), of masses of compressible material; that condensations in their arms were likely to be of masses comparable with the masses of the stars of our own galaxy; and that binary stars, or some of them, had probably been formed by a mode of fissional break-up, although wide binaries might be the consequences of capture of one nucleus by an adjacent one, only close binaries being formed by actual fission. It is not possible to reproduce all the details of Jeans's argumentation here, and it would scarcely be illuminating to do so, since Jeans was driven to estimates of mean densities of primitive gas-clouds which were largely guess-work. A discussion based on conjectural orders of magnitude can hardly hope to be of permanent interest. Nevertheless, an example of Jeans's prescience may be noted. He was writing in the days before it was certain that our own galaxy is one amongst many 'island universes' wholly outside it, and before it was realized that this universe of universes was in a state of expansion and mutual separation. Also Van Maanen's too short estimates of the rotational periods of the nebulae had not been disproved. But on p. 261 he forecast the phenomenon of the expansion of the universe when he remarked:

> We have already had occasion to contemplate a past epoch in the history of the universe, in which the stars were much closer together than they now are. We have found reasons for supposing that at this time the stars were close enough to affect one another's orbits in space, to an appreciable degree....

Jeans had concluded also that two of the other uniformities of structure could be accounted for by the effects of rotation, namely the planetary and ring nebulae, and the globular clusters. The former view is not now held, as it is more likely that the ring nebulae were formed by a process analogous to the occurrence of a nova, in which a sudden internal generation of energy, possibly occasioned by the collapse of a star on itself, results in the driving off of the outer layers by radiation pressure, and subsequent formation of a widely extended tenuous atmosphere. Jeans was on surer ground in his discussion of the possible formation of globular clusters of stars out of a rotating nebula, since these have since been observed by Hubble in the exterior parts of the Andromeda nebula. Jeans also discussed the probability of a state of equipartition of energy amongst the members of a globular cluster, owing to the enhanced frequency of encounters due to the high density.

But for the fifth uniformity of structure Jeans found no place in the rotational scheme of evolution. It was true that he had found that possibly planets might form out of the atmosphere thrown off equatorially from a rotating mass of gas, but there were several objections to this, primarily the objection that the next stage in evolution ought to be for the central mass to break up into an ordinary binary star, whereas our sun and planets are not binary. Also the arrangements of the components of multiple star systems such as might have been formed by rotation does not in the least resemble that observed in the solar system.

These were not the only considerations that threw doubt on the rotational theory of the origin of the solar system. It had been objected by Babinet in 1861 that the total angular momentum contained within the solar system appeared to be insufficient for the system ever to have broken up by rotation. Jeans therefore made a very detailed calculation of the probable available amount of angular momentum in the solar system, and concluded, not that it was inadequate, but that if ever it had been sufficient to cause break-up by rotation, the radius of gyration of the primitive sun must have been very small. This in turn implied that the primitive sun must have had a high degree of central condensation. When Jeans wrote, stars were not considered likely to have such a high degree of central condensation, but in my opinion, based on my own rerearches, the objection has much less force today. Nevertheless, further consideration showed that the ejected matter would necessarily have had a density some 200 times that of the outer parts of the primitive sun, and that this factor led to further difficulties when the subsequent evolution of the planets was considered. For on Laplace's nebular hypothesis, the planets threw off the satellites because further contraction caused increased rotation, but this would require impossibly large decreases in rotational periods; moreover the subsequent career of the planets would be one of fission into fragments of comparable mass instead of into one primary and several lesser satellites. Jeans concluded: 'We have conjectured that spiral nebulae, star clusters, binary and then multiple stars are formed by rotation; these complete the chain of rotational evolution, and there appears to be no room on this chain for systems like our own.' Jeans therefore turned to what is called the tidal theory of the origin of the solar system: 'The general conception of the Tidal Theory as applied to our solar system is that a second mass has at some past period approached so close to our sun as to break it up by intense tidal forces into a number of detached masses.'

His first argument in favour of this theory was that it would tend to give rise to a number of separate masses becoming detached from the primary mass and finally describing orbits about it, as against the result of the rotational theory, which could only lead, in the case of our sun, to a binary or at most a triple or multiple system quite unlike the system of solar planets.

The next argument is that the rotational theory fails to account for a difference of some 6° between the equatorial plane of the sun's present rotation and the invariable plane of the solar system: 'The tidal theory explains this naturally by supposing that the present invariable plane records the plane of passage of the tide-generating mass, whilst the present plane of the sun's rotation coincides approximately with that of the rotation of the original mass.'

Jeans then calculated that if the primitive sun just filled

the present orbit of Neptune, the critical distance for a mass double that of the sun to be capable of disrupting the sun would be just under three times the radius of Neptune's orbit.

He then calculated the probability that such a close encounter or passage would have occurred in the past. He concluded that the average interval between close encounters of the type indicated would, on the present density of stars in the galactic system, exceed the probable age of the universe, and so such an encounter would be an excessively rare event. The odds against the chance that the sun had been broken up in this fashion would be so large that the tidal theory would have to be discarded.

He pointed out, however, that it had previously been necessary to contemplate an earlier epoch in the evolution of the universe when the stars were much closer together than now, and the stars themselves more diffuse. Revising his estimate to suit these conditions, he concluded that the chance was increased 10,000 times. 'Tidal break-up, even now, can hardly be considered a likely event, but it is considerably more probable than our former calculations would have shown it to be, and the improbability of close encounters among the stars no longer provides adequate grounds for rejecting the tidal theory.'

The system is therefore likely to have come into existence when the sun was of low density. Taking the density of the flung-out nebulous arm as one-tenth of the mean density of the sun when extended to the orbit of Neptune, he calculated the size of the condensations likely to be formed owing to the longitudinal instability of the nebulous arm as equal to a mass intermediate between those of Jupiter and Saturn: 'It is clear that if our system contained, beyond the central sun, only planets of masses of the order of those of the two greatest planets, the tidal theory would provide a highly satisfactory explanation of the genesis of the system.'

But, 'to inspire confidence', it was necessary that the tidal theory should be able to account for the small planets as well as for the large planets, and for the satellites of the planets also. Direct considerations show (the arguments had been already advanced by Jeffreys) that the satellites of such a system as Saturn's, and probably of the other planets, are so small that they must have been solid or liquid from birth, as otherwise they must have disappeared by evaporation. It is difficult to suppose that solid or liquid satellites were born out of a gaseous planet. In the case of the earth-moon system, it was not possible to say whether the disparity of mass was too large for the system to have originated from a wholly fluid earth. But Jeans concludes that the earth was at birth more largely fluid than the planets with relatively smaller satellites.

To reconcile the conflicting considerations of liquid planets being born from a gaseous sun, Jeans examined the process of ejection of a jet of matter from the sun by a passing star in more detail. He had previously shown that the encounter would be likely to be of the 'slow' type:

> the rate of ejection of matter would be slow at first, and increase to a maximum when the passing star was at its distance of closest approach, and subsequently diminish to zero. The result ought to be a filament of matter of which the line density would be zero at each end and would increase to a maximum near the middle. As this filament lost heat by radiation, the ends would experience the greatest fall of temperature....Thus liquefaction ought to commence near the ends, and after a time the ends of the filament might be mainly liquid whilst the middle region was still almost entirely gaseous. During this process of condensation gravitational instability would result in the formation of furrows, leading to ultimate fission into separate masses.

The planets formed near the ends of the filament, being formed out of dense matter, would be those of smallest mass, whilst those formed near the middle, mainly from uncondensed gas, would be of greatest mass.

In this way the tidal theory readily explains the great inequality between the masses of Jupiter, Saturn and the other planets, whilst explaining at the same time why the two largest planets occur near the middle of the chain. The same theory indicates that the smaller planets must have been mainly liquid or solid from their birth, while Jupiter and perhaps also Saturn may have always been entirely gaseous.

After the filamentous arm had been ejected and the planets separated out, they would tend to follow orbits which would lead them back very close to the sun, though they would escape actual collision. The sun in turn would exert fresh tidal forces on the primitive planets during these close encounters, and this would result in the formation of systems of satellites like miniature solar systems. 'This would account for the directions of revolution of the majority of the satellites and explain why their orbital planes are, for the most part, close to the orbital planes of the corresponding planets.' It was also pointed out that the action of the resisting medium, which presumably must have surrounded the primitive sun immediately after the passage of the tide-raising star, would be to diminish the eccentricities of the planetary orbits, especially the inner ones, in agreement with the arguments of other writers.

Jeans concluded his sketch of the tidal theory (which of course he made no claim to have originated) by saying that he hoped it would be read rather as an indication of possibilities than as an attempt to advocate the theory or present it in a final form. Nevertheless, to Jeans it appeared infinitely more acceptable than the rotational theory, as an explanation of the genesis of the solar system.

He concluded the complete essay by denying that it had been part of his task to arrive at a conclusion: 'the time for arriving at conclusions in cosmogony has not yet come.' Yet he felt entitled to summarize the probabilities as follows:

Some hundreds of millions of years ago, all the stars within our galactic universe formed a single mass of tenuous gas in slow rotation. As imagined by Laplace in another context, this mass slowly contracted, conserving its angular momentum and so increasing its angular velocity. It assumed a lens-shaped form, then developed a sharp equatorial edge and threw off matter in the form of arms which took the forms of spirals. The long filaments of matter forming the arms were longitudinally unstable, and developed into chains of condensation-nuclei. This nuclei separated themselves, contracted and evolved into stars of incandescent material. As they shrank still further, they broke up by fission (many of them) forming binary stars.

Jeans thought that the original motions along the nebular arms persisted in the form of star-streaming; but, thanks to the work of Lindblad, it is now believed that star-streaming is a phenomenon of perturbations from the ideal circular or (in the present author's theory) spiral orbits. Jeans thought that stars or small groups of stars failed to get clear of one another's gravitational influences and formed multiple systems of wide separation. The more massive condensations persisted in the original equatorial plane of the nebula, and form what we now recognize as stars of spectral class B.

But at relatively very rare intervals, two stars passed quite close together. Some 300 million years ago (one would say today 3000 million years ago) our sun experienced a close 'encounter' of this type, a large star passing within a diameter of the sun's surface. This caused ejection of a stream of matter from the sun, which condensed into liquid near its ends, forming the small inner planets and the small outer ones, and forming the massive, possibly gaseous, planets near its middle. These condensations, owing to the orbital velocity communicated to them by the passing star, would not drop back into the primitive sun, but would pass the surface near enough for fresh tidal arms to be ejected from them, and these would in turn, by a similar process, yield the satellites. This at least accounts for the satellites of Jupiter and Saturn, though those of Neptune, Uranus, Mars and Earth are less easy to explain. The earth-moon system is the least easy to account for. The subsequent discovery of the small trans-Neptunian planet, Pluto, is in accordance with the general implication of the tidal theory. But the solar system, which includes the unique earth-moon system, may itself be unique in the system of the stars.

I have not attempted to bring the theory of the formation of the solar system up to date. The tidal theory, which originated as the planetesimal theory of Chamberlin and Moulton, and was much developed by Jeffreys, was subjected to searching criticism in 1935 by H. N. Russell, in his book *The Solar System and its Origin.* And since then Lyttleton has developed the theory that the sun was originally a binary or multiple system, and he and Hoyle have developed an accretion theory of the origin of the sun and stars. But Jeans never claimed finality for his conclusions, as we have seen. It is sufficient to stand aside from the details of controversy about the origin of the stars and solar system, and to admire the faultless way in which Jeans developed the solutions of the underlying mathematical problems, and completed the work of his great predecessors, Newton, Maclaurin, Jacobi, Kelvin, Poincaré, Schwarzschild and Darwin.

CHAPTER X

STAR CLUSTERS

In December, 1913, Jeans wrote an important paper in what was to him a new, though at the same time familiar, field. It was entitled 'The kinetic theory of star clusters'. It was his first paper in the *Monthly Notices* of the Royal Astronomical Society. The field was familiar, because it made a calculation similar to those Jeans had been frequently making in his work on the kinetic theory of gases; it was new, because it was the first occasion on which Jeans ventured into the theory of star clusters. Its object was to determine the deviation of direction in the motion of stars due to close approaches by other stars.

The formula I need not quote. But some of Jeans's applications of it are of interest. Assuming 10^9 stars within a distance of 1000 parsecs,* assuming each star to be on the average five times as massive as the sun, and assuming a typical relative velocity between members of a pair of stars 'in collision' to be 60 km. sec.$^{-1}$, he calculated that a gross deflexion of 1° would be acquired after a distance of 4×10^{23} cm., or, with an individual star velocity of 40 km. sec.$^{-1}$, after an interval of 3200 million years, which is the present estimate of the age of the universe. This is the result of the accumulation of small deflexions, excluding 'violent' deviations. A sudden deviation of, say, 5° would occur once in 5 million million years; and one of 2°, once in 8×10^{11} years.

He further summarized what might be deduced from his formula in the following vivid passage:

...let us take a definite instance of a star stream in which the stars all start with equal and parallel velocities of 40 km. sec.$^{-1}$.

* A density far in excess of the density of the universe accepted at present, but suggestive of densities in star-clusters.

Let us suppose that a star is still considered to belong to the main stream as long as its direction of motion makes an angle not greater than 2° with the main stream. After 100 million years, the stream will have lost only one in 8000 of its original members, and the remainder will make angles with the main stream of which the average amount is only 10′. After 3200 million years the loss will be one in 250 and the average angle will be 1°. After 80,000 million years, one-tenth of the original members have been lost by violent encounters, but the average angle of the remainder is 5°. Thus the stream has been mainly dissolved, not by collisions or violent encounters, but by gradual scattering.

Jeans concluded from this that there was no likelihood of our universe (he meant galaxy) being in a steady state. This consideration led him to tackle, in 1915, the general problem of the distribution of velocities in a stellar system. He first investigated the general conditions under which Schwarzschild's ellipsoidal law of dispersion in stellar velocities could prevail. He made what is probably the first application of Boltzmann's equation in gas theory to the case of stars, and set the pattern for future writers on this subject. His own conclusion was that, except for 'very artificially constructed universes', the general result always emerges that for Schwarzschild's law to be consistent with a steady state, the universe must be either spherical or spheroidal, excepting in the special case in which the Schwarzschild ellipsoid reduces to a sphere, so that no star-streaming occurs at any point of the universe. He went on:

It may probably be taken for granted that our universe is neither spherical nor of the special spheroidal type we have had under consideration. It may be taken for certain that star-streaming occurs in our own universe, at least at our particular instant of time and in our particular region of space.... We are led to the conclusion that our universe cannot be in a steady state.

He foresaw that this conclusion would not be readily accepted. He therefore examined the general problem by a more powerful mathematical method, which again has often been followed by later writers. By comparing his kinematics with the dynamical conditions necessary for it to exist, he again concluded that the system was not in a steady state. In a passage which foreshadowed his future success as a popular expositor, he illustrated his conclusion as follows:

It ought perhaps to be noticed that there was never any valid reason for making the assumption that the universe is in a steady state. Indeed, the *a priori* probabilities are all in the other direction. The direct observational knowledge which we have of the movements in our universe is excessively limited both in time and space: the introduction of the steady-state hypothesis is only a tentative effort to remove the restriction in time. In the same way a being whose life was infinitely more restricted than ours might try if he could explain such phenomena as an earthquake or a lightning flash in terms of a steady-state hypothesis. His attempts, as we know, would be foredoomed to failure, and the results we have obtained seem to indicate that attempts to explain star-streaming as a steady-state phenomenon must be equally foredoomed to failure.

This pessimism did not prevent Jeans from making a further attempt on the problem in 1916. In this paper he worked out the details of the 'collision' or interpenetration of two star clusters in relative motion. He was again very diffident about his results:

In conclusion it must again be repeated that the problem we have had under discussion has been far too complex for anything like a full solution to be obtained. In our brief investigation it has been repeatedly forced upon us that there must be many agencies at work simultaneously in the process we have been considering, while we have confined attention to one. It is as though a rich pattern were being woven, and we had been able to follow one thread. The question at issue is whether, by following this one thread, we have been able to form any conception, even the most incomplete, of the weaving of the whole pattern.... I am inclined to suggest, although only tentatively and conjecturally, that this one thread may really provide a clue to some of the processes which are at work in the weaving of the whole.

Lastly, Jeans followed this with a short paper on 'The law of distribution in star clusters'.

It must be remembered that these were before the days of Lindblad's theoretical and Oort's observational researches on the rotation of the galaxy, confirmed spectroscopically by J. S. Plaskett. Star-streaming is now tolerably well understood. The velocity dispersion observed in star-streaming was shown by Lindblad to consist of perturbations of purely rotational motion round the centre of the galaxy: stars in the vicinity of the sun will not all be describing strictly circular orbits round the galactic centre, and their motions, relative to a star carried round with the galactic rotation will give rise to the phenomenon of star-streaming according to the Schwarzschild velocity-ellipsoid. I have shown that similar results follow according to kinematic relativity, when the principal circular orbits are replaced by spirals. Jeans's conclusions were thus not destined to last. But the pioneer mathematical methods which he used have been of the greatest use to his followers in this field, such as S. Chandrasekhar, G. L. Camm and Shiveshwarkhar. As so often, the fundamental mathematical background is more permanent than the immediate applications in the light of the observational facts then available.

CHAPTER XI

THE EQUILIBRIUM OF THE STARS

In 1928 Jeans published his *Astronomy and Cosmogony*, which he considered to be in a sense a sequel to his *Problems of Cosmogony and Stellar Dynamics* of a decade previous. Actually it comprises much that was in the Adams Prize Essay; but it also includes much original work, besides the contents of a long series of papers in the *Monthly Notices* of the Royal Astronomical Society. Though this book is exciting and interesting from cover to cover, it cannot be regarded as a masterpiece of the quality of the Adams Prize Essay, for reasons stated earlier. Nevertheless, the ideas on stellar structure contained in this book are well worth recapitulating.

He begins with a survey of the facts of observational astronomy, with an account of the distances of the principal types of astronomical objects, and catalogues in turn the properties of binary stars, variable stars, triple and multiple systems, moving clusters of stars, globular clusters, planetary nebulae, irregular nebulae and extra-galactic nebulae. This leads him to ask the fundamental questions:*

What, in ultimate fact, are the stars? What causes them to shine, and for how long can they continue thus to shine? Why are binary and multiple stars such frequent objects in the sky, and how have they come into being? What is the significance of the characteristic flattened shape of the galactic system, and why do some of its stars move in clusters, like shoals of fish, while others pursue independent courses? What is the significance of the extra-galactic nebulae, which appear at a first glance to be other universes outside our own galactic universe comparable in size with it, although different in general quality? and behind

* *Astronomy and Cosmogony* (1929 ed.), p. 29.

all looms the fundamental question: What changes are taking place in this complex system of astronomical bodies, how did they start and how will they end?

Many of these questions were answered in the Adams Prize Essay, and *Astronomy and Cosmogony* does not carry this part of the story much further. But here we shall try to give an account of Jeans's ideas 'of the object which occurs most frequently of all in nature's astronomical museum, the simple star'.

His first chapter on this subject gives an admirable account of stellar magnitudes and luminosities, colour indices, bolometric correction, stellar spectra, spectral classification, effective temperatures, the characteristics of typical individual stars, the temperature-luminosity diagram (Hertzsprung-Russell diagram) and the white dwarfs.

His next chapter is devoted to his theory of gaseous stars. At the beginning of his investigation he rightly puts Lane's law—the law that for homologous equilibrium configurations, the temperature at any point is inversely as the radius of the configuration. He then proves the paradoxical but well-known result that if κ (ratio of specific heats) exceeds 4/3, a loss of energy will cause the star to contract, but at the same time increase its temperature so that the more the star radiates heat the hotter it becomes. He next proves the theorem of the Virial—one of those powerful global relations like the conservation of mass and of energy which give information of amazingly precise nature without going into details.

This leads him to an account of adiabatic equilibrium in a star and its generalization in the form of Emden's polytropic spheres. With $\kappa = 1\frac{2}{5}$, he then calculated from Emden's formulae the following figures for an adiabatically arranged sun built up of gas of the molecular weight of atmospheric air:

Temperature at centre of sun $= 455 \times 10^6$ degrees.

Density at centre of sun $= 34{\cdot}02$ gram cm.$^{-3}$.

Pressure at centre of sun $= 43 \times 10^9$ atmospheres.

But he then pointed out, as he had been the first to point out in 1917, that this ignored the effects of the high temperature on ionization. At temperatures even much lower than the figure just computed, molecules and atoms would largely be broken down into free electrons and bare nuclei, and they would each separately make a contribution to the pressure. For example, a nitrogen molecule of atomic weight 28 would be broken down into two nuclei of mass 14 and fourteen free electrons. This would, for a given temperature, increase the pressure 16-fold; or (in the form relative to the present problem where the total pressure is conditioned by the mass supported), for a given pressure the temperature would be decreased 16-fold. The mean mass of the particles would be of the order of 2·0, and this would be so, approximately, whatever the chemical element considered. With mean particle-mass equal to 2, the computed temperature at the sun's centre is reduced to

31·5 million degrees.

For the stars in general, he went on later to argue that they would be composed of elements of atomic number equal to or greater than that of uranium (the heaviest atom then known), and such very heavy atoms would not be completely broken up by ionization. They would retain their inner, most tightly bound electrons. Taking this into account he inclined to a mean particle-weight of about 2·6, and this revised estimate sends up T_c, the central temperature, into the region of 40 million degrees. This calculation assumed that the material at these high temperatures and pressures obeyed the laws of these gases. He foreshadowed that he would soon come on grounds for doubting this assumption. The calculation also ignored the mechanical effect of the pressure of the radiation that was surging out from the star. Discussing this, he concluded that radiation pressure would be equivalent to reducing the sun's mean molecular weight at its centre by some 17 per cent. He also stated that the effects of radiation pressure became considerable in stars whose mass was several times that of the sun.

He next turned his attention to the details of the mode of transfer of energy by radiation. He showed that transfer of heat by conduction, according to well-known formulae of the kinetic theory of gases, would be quite inadequate to supply the vast quantities of energy observed to be leaving the sun's surface. Arguing by analogy with the dynamical theory of gases, and taking a mean free path-length of radiation as defined in terms of the coefficient of absorption of radiation, he was able to make a rough calculation of the coefficient of 'conduction' when the mode of transfer was by radiation, and to show that it enormously exceeded the ordinary coefficient of gaseous conduction. In these calculations he was covering ground explored earlier by Chapman and by Eddington.

When he went into details concerning the flow of radiation, he accused 'Eddington and many of those who followed him' of being led into a series of errors through not sufficiently recognizing the approximate nature of their equations. This was, to say the least, an *ex parte* statement arising out of Jeans's former disagreement with Eddington. Actually Eddington had fully attended to the degree of approximation of his formulae, and shown that under stellar conditions they held good with close accuracy. There were, it is true, perversities in Eddington's analysis which he was loth to admit, but when they were corrected there was no appreciable difference in the resulting formulae. The investigation which Jeans here quotes from a paper of his (Jeans's) in *Monthly Notices* ((1926), **86**, 576) had in fact been almost wholly anticipated by a paper of mine in the *Proceedings of the Cambridge Philosophical Society* ((1923), **21**, 701), as Jeans indeed there stated. When there were serious points of substance on which Eddington's theory of internal constitution of the stars could be criticized, fault-finding in this particular matter can only be called a storm in a teacup.

Jeans now incorporated the theory of radiative equilibrium into the theory of polytropic gas-spheres of general index n, and though his method was not rigorous he rightly showed that if n was unequal to 3, a star of given parameter could always find a position of equilibrium by adjusting its central density (or radius). But as n reaches the value 3, 'the addition or subtraction of the slightest amount of mass causes the star to rush through the whole range of values from $\rho_c = 0$ to $\rho_c = \infty$'.* This led him to his fundamentally sound criticism of Eddington's theory as being based on $n = 3$, according to which, on Eddington's own showing, the luminosity depended on the mass only, independent of the nature of the energy sources.

Jeans next discussed the physical formula for the absorption coefficient of stellar material, and, with some generalizations, investigated the behaviour of the ratio of radiation-pressure to gas pressure near the surface and in the interior of a star, concluding, in agreement with other workers, that this ratio tends in the interim to some constant value, or slowly varying value, depending amongst other things on the rate of generation of energy per unit mass. He now combined this formula with the equations of dynamical equilibrium.

But he proceeded to derive from earlier passages, a statement which contradicted his own criticisms of Eddington's theory. He states: 'For stars of mass considerably less than that of the sun, λ [the ratio of gas pressure to radiation pressure] will be quite large', forgetting that he has just shown that it can have any value, depending on the rate of generation of energy. However, it was in this section that he drew attention to the fact that the polytrope, $n = 3\frac{1}{4}$, suits the constitution of a star of constant molecular weight and uniformly distributed energy sources, with a reasonable formula for the opacity—of course, on the perfect gas hypothesis. With stars of large mass he encountered a difficulty. He then devised an approximate method of reducing the

* *Astronomy and Cosmogony*, p. 82.

stars of intermediate mass to suitable polytropes, and applied it to derive formulae for the pressure and temperature at the centres of stars.

This led him to a formula connecting N^2/A (where N is the atomic number of the material of which the star is composed and A is the atomic weight) with the central temperature, the mean rate of generation of energy, and the central value of the ratio of gas pressure to radiation pressure. He attached great importance to this formula, but it is a very odd sort of formula, as it does not express N^2/A in terms of quantities immediately given by observation, but in terms of central temperature and relative importance of radiation pressure at the centre, which he has already calculated by separate formulae. It is difficult to see directly the physical meaning of the formula. Jeans interpreted it as stating that 'in gaseous stars which have the same given value for M [mean molecular weight], μ [mass] and N^2/A—i.e. stars whose mass and composition is fixed—T_c [the central temperature] is proportional to the square of the luminosity of the star'.* And since the product of central temperature and radius of a star is constant for stars of given mass, it followed that 'the radius of the star is inversely proportional to the square of its luminosity'. These, he said, were the laws which would be obeyed (on the simplifying approximation made), if the gas-laws were accurately obeyed throughout stellar interiors. But he also said that observational astronomy gave no support to such a law. It is difficult to see what Jeans meant by this, since, for a star of given mass, the luminosity is fixed, and therefore there is no scope for investigating the variation of radius with luminosity for given mass.

However, Jeans went on to attempt to use his formula to determine the atomic weight of stellar matter, on the supposition that the gas-laws are obeyed. His results appeared to indicate atomic weights far higher than that of uranium (the element with the heaviest atom then known), and in

* *Astronomy and Cosmogony*, p. 101.

most cases atomic weights of thousands at least. He pointed out that there was nothing fundamentally impossible in this conclusion, since the atoms which chemists have studied have presumably been derived from the outside regions of stars. But there was one fundamental objection to the conclusion on the score of self-consistency; for it implied a value of the mean *molecular* weight (allowing for ionization) greater than had been presupposed in the calculation, and if a higher value were adopted, still higher values of N^2/A, and so of the atomic weight, would result.

He then considered the effect of changing the stellar model assumed, in the sense of concentrating the region of energy generation towards the centre. This afforded some relief from the dilemma, but not enough; nor did it help much to assume a steep gradient of atomic weight towards the centre. He was driven to consider a more drastic possibility: the possibility that all the interior temperatures were overestimated. Reducing T by something less than a half, he showed, would lead to reasonable values of the atomic weights.

But the pressures in a stellar interior are fixed as regards order of magnitude by the circumstance that they have to support the mass against gravity. Hence, the mean density being given by the observed mass and radius, there is no scope for altering the temperatures so long as the ideal gas laws are obeyed.

Suppose then that the temperatures are artificially reduced. The pressure, both the gas pressure and, still more, the radiation pressure, would be reduced and the star would tend to collapse. But the diminished temperatures could not maintain the previous degree of ionization, some recombination of previously bare nuclei with orbital electrons would occur (e.g. by the formation of 'K-rings') and the ionized atoms so formed could no longer be treated as points; the gas laws would then not be strictly obeyed. His ultimate conclusion would be that stars are composed of atoms of about atomic number 95, and he computed that a reduction of temperature by 30 per cent would be sufficient to yield values of N^2/A corresponding to this. But he computed further that the ionized atoms would then be so closely in contact that their effective volumes would occupy all the available space. 'Such a condition may be properly described as a liquid, or semi-liquid, state.'*

To summarize,

> if Boyle's law is assumed to be obeyed throughout the interior of a star, the observed capacity of the atoms for stopping radiation demands an impossibly high atomic weight. We can reconcile the observed opacity with reasonable values of the atomic weights by supposing the density to be so great that Boyle's law is not obeyed, but the deviations from Boyle's law have to be so great that the matter must be supposed to be in a liquid or semi-liquid state.

He foreshadowed that two independent lines of evidence would lead to the same conclusion.

The next chapter in his book discusses the source of stellar energy. It recapitulates the arguments from orders of magnitude which show that the heat content stored in a star is inadequate, and that the gravitational energy is inadequate. This leads on to sub-atomic sources of energy, and to Einstein's relation between mass and energy. Jeans rightly pointed out that when mass is annihilated, an equivalent amount of energy is liberated, and he emphasized that the amounts of energy available in such processes are enormous compared with the amounts in any chemical process. (It must be remembered that this was a decade and a half before the atomic bomb.) As regards the particular process by which mass would be destroyed and energy created, Jeans favoured the idea of the mutual annihilation of a proton and an electron. In connexion with this speculation, he quoted an interesting passage from Newton's *Opticks*, to which his attention was drawn by

* *Astronomy and Cosmogony*, p. 108.

Sir J. J. Thomson; this passage ends with the query: 'And amongst such various and strange transmutations, why may not Nature change bodies into light and light into bodies?' This suggestion of Jeans was not destined to be the process which fifteen years or so later was thought to be the actual process of generation of stellar energy, namely the carbon-nitrogen cycle, by which four protons eventually unite to form a helium nucleus. But the investigation which his suggestion led him to prosecute turned out to be both fascinating and fundamental. It was no less than a full-scale investigation into the stability of a gaseous star. This we now recapitulate.

Just as two differential equations determine the equilibrium of a star, namely an equation of mechanical equilibrium and an equation of transfer of energy, so two equations determine the dynamical motion of a star not in equilibrium, namely an equation of motion and an equation expressing the rate of change of temperature with time at any point. The latter equation was first correctly formulated by Vogt in 1928, as acknowledged by Jeans in a footnote. Jeans now generalized the analysis to a certain extent by introducing a modified equation of state of a gas, containing an exponent S which vanishes for an ideal gas. Further, he introduced two exponents α and β such that the generation of energy per unit mass in the star's interior was proportional to $\rho^{\alpha}T^{\beta}$, ρ and T being density and temperature. He then eliminated δT, the variation of temperature from its equilibrium value, between the two equations determining the dynamical motion already referred to, and was thus led to a differential equation of the third order in δr, the variation of radius, as a function of time t. He then discussed the roots of the corresponding cubic equation, assuming the coefficients in his differential equation to be constants.* The stability of the star turns

* He had assumed δr proportional to r throughout the volume of the star, so that 'changes in any one shell are characteristic of the whole star'.

on the reality and signs of these roots. He obtained two conditions for stability, involving chiefly the exponents α, β and s, n the exponent in the opacity-law and λ the ratio of gas pressure to radiation pressure. The one inequality expressed the condition that small pulsations, of a few hours or days period, should not increase indefinitely in amplitude, the other expressed the condition that there should be no unstable expansion of an explosive type. These conditions involved limitations on the values of $3\alpha+\beta-n$.

In his endeavour to simplify these conditions Jeans was led into one of those obscure patches of mathematics in this volume which his most fervent admirers can only deplore. Having catered for departures from the laws of a perfect gas by putting pressure proportional to temperature times density to the power $(1+s)$, he now made another and algebraically incompatible mode of departure from the gas laws by putting the pressure equal to $(1+\xi)$ times the value it would have for a perfect gas. For a perfect gas he has now the two conditions $s=0$, $\xi=0$. However, he was not abashed by this, but proceeded to show that for stars of very large mass, the conditions for stability boil down to $3\alpha+\beta-n=0$ for a *gaseous* star. He concluded that it was 'infinitely improbable' that stars of very great mass, now observed to be existing stably, could be wholly gaseous. 'If $3\alpha+\beta$ were greater than about unity, Pearce's star, and probably also the more massive stars 27 Canis Majoris and Plaskett's star, would be unstable through a tendency to develop explosive pulsations.'*

He next considered various possible physical mechanisms for the generation of energy; he concluded that interactions between atoms, or between atoms and free electrons, or between atoms and radiation, could all be ruled out on the grounds that they would violate his stability criterion. He even argued that the synthesis of helium from four protons would proceed at such a rate as to give explosive instability,

* *Astronomy and Cosmogony*, p. 125.

but, as remarked earlier, this was before the role of the carbon nucleus as a catalyser was realized. Lastly, he ruled out a suggestion of H. N. Russell's that the generating stellar energy is negligible until the temperature attains a certain critical value. By elimination of these possibilities, he concluded that the generation of energy must be a monatomic process, and hence independent of temperature and density; in fact, that it must be akin to radioactivity, which theoretically requires a temperature of $7{\cdot}5 \times 10^{12}$ degrees to influence it. 'What is annihilating the matter of the stars is neither heat nor cold, neither high density nor low, but merely the passage of time.'*

An inference from this was that the stars which are radiating the most intensely are not the stars of highest or lowest internal temperature, or highest or lowest density, but the youngest stars, and so the most massive stars. Consideration of the actual rates of radiation by actual stars, from the most massive observed to the humble red dwarf, confirmed this.

Jeans now adopted the simple empirical formula $L \propto M^3$ as giving the dependence of absolute luminosity on mass and this, combined with the Einstein formula $dM/dt = -L/c^2$, allowed him to make an estimate of the ages of the stars. He concluded that they were of the order of 10^{12} years to 10^{13} years. It is curious, and unsatisfactory, that after all his attempts at calculating the rate of loss of energy from a star in a given physical state, Jeans's final step was to assume a purely empirical formula for the dependence of luminosity on mass.

He concluded this chapter with a vigorous pen-picture of the mechanism by which, as he assumed, the heavy radioactive atoms he postulated for the sun's interior actually generated energy. He conjectured that one of the bound electrons fell into the nucleus, and so annihilated a proton, generating thereby a photon of extremely penetrating

* *Astronomy and Cosmogony*, p. 128.

radiation, which, after many scatterings, absorption and re-emission was degraded into photons of lower frequency and was finally ejected from the star's surface.

It may be remarked that the 'long' time-scale of 10^{12} to 10^{13} years is incompatible with the more securely based 'short' time-scale of about 2×10^9 or 3×10^9 years, deduced from the phenomenon of the recession of the galaxies; and that as the proportions of helium to lead in rocks in the earth's surface indicate a similar age, it is not now supposed that the stars have radiated away any appreciable proportion of their masses since they came into existence.

In Chapter v, Jeans now turned his attention to what he called 'liquid' stars, i.e. stars in whose interiors the ideal gas laws are not obeyed. At the beginning of the chapter he gave a new formula for the internal temperature of any star in terms of rate of generation of energy, mass and radius, though of course only one of these is an independent variable when the law of generation of energy is fixed. But it led him to a conclusion (which is not in agreement with the later researches of others) concerning the central temperatures of white dwarf stars. These, he concluded, must be enormously high: but more recent researches have shown that the only temperature gradients in a white dwarf star occur in the gaseous fringe, the degenerate interior being approximately isothermal in virtue of its small opacity. This means that the internal temperatures of white dwarf stars are comparatively low. Jeans, of course, was writing before the theory of degenerate gases had attracted attention. In the same sentence Jeans concluded also that the central temperatures of giant stars of large radius must be comparatively low. This would have led to the paradoxical conclusion that the hottest interiors are the lowest generators of energy, and the coolest the highest. Actually the reverse is the case. White dwarfs can scarcely exceed some 15 million degrees in their roughly isothermal central regions, as against Jeans's suggestion that their temperatures must be measured in some thousands of millions of degrees. Red giants, on the other hand, probably possess a small, very hot core, much hotter than the 20–40 million degrees usually attributed to the centres of ordinary stars of the main sequence.*

With this picture of a stellar interior before him, Jeans returned to the question of the stability of the stars. Emphasizing his former argument that the expression $3\alpha+\beta$ must be approximately zero,† Jeans now neglected α and β and took $n=\frac{1}{2}$ (the exponent in the law of opacity), and derived as the condition for stability

$$s>\frac{1}{45}\left(1+\frac{4}{\lambda}\right),$$

where λ as before is the ratio of gas pressure to radiation pressure. 'For stars of moderate mass, λ is large, and the condition for stability is that s must be greater than 1/45; for stars of large mass λ is comparatively small, and s must have a value substantially higher than 1/45.'‡ It will be remembered that the exponent s was Jeans's measure of the departure of the material from the state of a perfect gas. He here pointed out that even the small values of s made enormous differences to the estimates of internal temperature of stars, and to the estimates of pressures.

He concluded that 'unless the atoms in the star's central regions are packed so close as to provide a firm unyielding base..., the star will be liable to start contracting or expanding, this contraction or expansion continuing unchecked until a firm base is formed at its centre'. 'This is what we may describe as a semi-fluid state.' Hence his term, 'liquid stars'.

* Cf. my letter to *Nature, Lond.* (12 Dec. 1931), on this subject.

† Where, as before, α, β are exponents in the law supposed to express the dependence of energy-generation on density and temperature, α and β both of course being zero for ordinary radioactivity except possibly at temperatures of the order of 10^{11}–10^{12} degrees.

‡ *Astronomy and Cosmogony*, p. 145.

Jeans now began a new investigation of stability on more general foundations. He showed in a simple way that, if E is the total emission of energy from the surface of the star and G the rate of internal generation, then a transition from dynamical stability to instability occurs whenever

$$\frac{d}{dR}(E-G)$$

passes through a zero value; for such configurations must be in *neutral* equilibrium; and he showed finally that it must be *positive* for dynamical stability. When G is independent of R (the radius) the deviations from the gas laws must be such as to reverse the sign of dE/dR. He stated that, to discuss this criterion, we need a knowledge of the deviations from the gas laws, and that this depends on a knowledge of the state of ionization of stellar matter. Unfortunately, when he came to discuss the degree of ionization, he used a formula (due to Lindemann and Saha and developed by R. H. Fowler) which is only valid for a perfect gas. However, he struggled on with these hybrid formulae, and eventually concluded that

as a star steadily contracts along a homologous series of equilibrium configurations, the deviations from Boyle's law in its central regions may fluctuate through a succession of maxima and minima as the different rings of electrons surrounding the atoms are ionized in turn. The contraction of the star is accompanied by an increase in its temperature and, by ionizing one ring after another of electrons, this causes the atoms to diminish in size. The star and its atoms contract together but the star contracts steadily while the atoms, so to speak, contract by jerks.

He considered that the M, L and K rings* of electrons were successively removed by the process of ionization; each time a ring became ionized there was a tendency to regain

* The reader should be aware that here M, L and K are taken from X-ray nomenclature; M and L do *not* here denote mass and luminosity.

perfect-gas conditions, but further contraction caused congestion of the atoms, and another ring of electrons was shed, until finally the atoms were ionized down to their bare nuclei. The regions in the graph of E against R divide themselves into regions of stability and regions of instability. When changes of G are taken into account, the ranges of stability become more restricted, and he attempted to show that the effective temperature-luminosity diagram (the Hertzsprung-Russell diagram) is divided into narrow zones of stability—the tenanted portions—separated by wide zones of instability which are untenanted. Thus the giant red stars correspond to the region of stability following complete M-ring ionization; there follows a gap caused by the ionization of the L-ring, and the resulting region of K-ring congestion corresponds to the main sequence of stars. To the left of this main sequence is a gap caused by the ionization of the K-rings of the atoms, and finally another narrow stable band occurs, corresponding to the white dwarfs.

Whatever one thinks of the details of the investigation, it was a great *tour de force* to arrive on these grounds at a reproduction of the Hertzsprung-Russell diagram that corresponded so well with observation. But I confess I am unable to follow physically the arguments by which Jeans claimed to show the successive degrees of ionization of the stellar atoms; his mathematical analysis is tied up with his erroneous formulae for the central temperatures of the stars. Thus, for example, he attributed the maximum possible degree of ionization to the white dwarf stars because he had concluded earlier that they were the hottest. But of course this faulty conclusion vitiates all the deductions he makes from it, and in fact Jeans's views on the successive ionizations of the electronic rings of the atoms have not been accepted.

Jeans proceeded to estimate the atomic numbers of the atoms whose successive ionizations he had predicted and by two independent but rather obscure calculations arrived at atomic numbers of the order of 93, thus confirming his view reached earlier that the atoms near the centres of the stars are the radioactive atoms of high atomic number. The actual calculation is done by fitting portions (or, rather, selected points) in the tenanted portions of the Hertzsprung-Russell diagram with his own predictions concerning ionization.

By a combination in effect of Lane's law with his own views on the successive ionizations, he found that on his theory the minima of ionization of the *L, M* and *N* rings ought to occur at temperatures which are approximately in the ratio of

$$\frac{1}{2^2}:\frac{1}{3^2}:\frac{1}{4^2}$$

or* 36 : 16 : 9. He admitted that, on his own calculations, the range of internal temperatures is wider than this would suggest, but he said that we need attach little weight to the discrepancy.

He summed up his investigations on stellar structure by saying that 'considerations of stability demand that all stars should be in a state in which the deviations from the gas laws are appreciable, while actually the majority are found to be in states in which these deviations are so large, that their central regions may properly be described as in the liquid state'. And he repeated that the configurations which his theory predicted to be stable, grouped themselves in respect of radius, luminosity and spectral type in the way in which the observed stars are found actually to be grouped, *M*, *L* and *K*-ring ionization corresponding to the three categories of the giant stars, main sequence stars and white dwarf stars. To secure agreement of these categories of stars with the quantitative theory, he found it necessary to assign atomic numbers to the stellar atoms of the order of 95. This of course fitted in with his previous conclusion that the origin of stellar energy was due to a type of super-radioactivity.

* Sic. *Astronomy and Cosmogony*, p. 163.

The type of radioactivity he envisaged was different from terrestrial radioactivity in that it involved the annihilation of mass and its replacement by an equivalent energy of radiation. Atoms of this type not being known on earth, they would necessarily have atomic numbers greater than 92, that of uranium, the heaviest atom then known.

How much of this theory survives today? As remarked above, the conclusions as to the ionization of successive rings of electrons and its correspondence with different categories of stars have not been accepted. But a good deal survives of Jeans's view that many of the stars are not composed of perfect gas. At a time when Eddington's theory of the structure of the stars was dominant, it required great courage on Jeans's part to put forward a theory so different, at so early a date as 1928. In 1929 I came to the conclusion that Eddington's arguments relating to the all-gaseous character of stars and his arguments purporting to derive the internal opacity of stars were fallacious, by a line of reasoning totally distinct from Jeans's. Jeans, indeed, followed Eddington in obtaining formulae which purported to give the internal opacity in terms of the mass, luminosity and effective temperatures of the stars, though his formulae differed considerably. But my view that, as we penetrate a star from the outside, we must eventually encounter a region where the laws of a perfect non-degenerate gas cease to hold is not so very far from Jeans's hypothesis of liquid stars. And today the view is widely held that most stars possess a core in convective equilibrium, i.e. in a different condition from that of the purely radiative equilibrium of Eddington's 'standard model'. The giant red stars are indeed often supposed to have a small, very dense core of non-perfect gas, and I think a conclusion of this type is not very far removed from Jeans's views. Of course, when the radiative model is not continued right through to the centre, the solution of the corresponding polytropic differential equation is not in general the one which corresponds to a complete polytrope without central singularity. Such singularity-possessing solutions, which were first considered by me in a stellar context and later were fully explored by R. H. Fowler, were not considered by Jeans, but in fact they pave the way for a small central core of high density not obeying the classical gas laws.

The general impression which Jeans's work on stellar equilibrium makes on the reader is that he had not proposed to himself sufficiently clear-cut questions couched in the mathematical language he was employing. The consequence is that the pace is forced in an unhealthy manner, and there is little of the monumental or sculptural quality about these investigations. There are few formulae amongst the many he deduced which are commonly quoted. The time was not ripe for a discovery of the reasons for the observed characteristics of the stars; indeed it is not ripe today. Writing before the theory of degenerate perfect gases was familiar to astronomers, and before the realization of the expansion of the universe put the weight of emphasis on the 'short' time-scale of a small multiple of 10^9 years, Jeans was unfortunate in his decade of researches in this subject. It is probable, however, that the narrow tenanted portions of the Hertzsprung-Russell diagram (luminosity plotted against spectral type or effective temperature) do correspond to states of stability, and the untenanted portions to regions of instability, as supposed by Jeans.

That Jeans was not entirely clear about the objectives in these studies on stellar equilibrium is exhibited in the next chapter, called 'The evolution of the stars'; for, one-third of the way through the chapter, he wrote: 'We must first find the relation between a star's mass, luminosity and surface temperature when the gas laws are supposed to be obeyed throughout the star's interior.'* It is a little odd that after so many pages of analysis of stellar equilibrium

**Astronomy and Cosmogony*, p. 172.

he should only at this stage have formulated to himself what for most writers has been the fundamental problem to attack. The formula he thus reached is akin to the similar formula due to Eddington. He made use of it, not as Eddington did to confront the numerical data of observation, but to trace the theoretical evolution of a star of diminishing mass. He concluded that, if a star consisted of a single type of material liberating energy at a constant rate per gram, then as its mass diminished it would pass through the red, orange, yellow and blue stages in succession, so that the hottest stars would be the least massive, contrary to what is found in nature. He therefore went on to consider a star, part of whose mass remained unaltered while the remainder evolved energy at a constant rate. He showed that, for such a star, the surface temperature would begin low, increase to a maximum, then diminish again to the red stage and ultimately lapse to invisibility. This evolutionary course would correspond to that originally envisaged by H. N. Russell, giving a path through the *M*-type giants to the *A*- and *B*- and possibly *O*-type stars, followed by the main sequence path *O*, *B*, *A*, *F*, *G*, *K*, *M*.

He then considered the influence of fission on a star's evolution. He deduced that, if a red, gaseous *M*-type star of effective temperature 3000° breaks up into two equal, similar halves, each component would have an effective temperature of the order of 18,000°, and so be of spectral type *B*2 or *B*3. He found confirmation of this in the circumstance that 'newly formed' binary stars are generally observed to be of spectral types *O*, *B* or *A*. If a star divided into two unequal parts, the less massive would have the higher effective temperature; and when the less massive component had decreased in mass to any given mass, it would be of greater luminosity and higher effective temperature than the other component when the latter had reached the same mass. This was on the assumption that the energy liberation of the two components was proportional to their masses; but this does not agree with observation, as Jeans pointed out. He therefore modified this hypothesis in the sense that, when a star breaks up by fission, the less massive is formed mainly from the outer parts of the parent star, on the assumption that the heavier, energy-generating atoms have been concentrated towards the centre of the parent star. This would imply that the more massive component would start life with a higher rate of liberation per gram than the parent star, and further considerations suggest that the disparity in value of energy liberation per gram for the two components would be greater in old binaries than in young ones. Some confirmatory evidence is contained in work by Leonard at the Lick observatory to the effect that in giant binaries the brighter component is usually of later spectral type than the fainter component, while in dwarf binaries the reverse holds good.

Jeans now pointed out that these conjectural courses of evolution would be changed if the stars were no longer supposed to be wholly gaseous. Instead of slow secular evolution, the regions of instability in the Hertzsprung-Russell diagram would be ephemerally occupied by stars in a rapid state of evolution. He claimed that now, if the effective temperature of a parent star was fairly high, both components would be thrown right over to the left-hand edge of the main sequence of stars, which he regarded as an impenetrable barrier, owing to the crowding together of the atoms, in the absence of the next stage of ionization. Their evolutionary sequence would then be a march down the main sequence. As regards the giant *F*, *G*, *K*, *M* stars, he inferred that when fission took place amongst them, the type would change suddenly to become *O*- or *B*-type. The same change might occur without fission, if a giant star came to the unstable edge of its region of stability, and it would then become a main-sequence star of early type.

Some main-sequence stars however, would break through the barrier, and become white dwarfs, in which the atoms (he supposed) would be stripped down to bare nuclei and be only feeble generators of energy. He even considered that the sun might be in a state in which it could collapse at any moment into a white dwarf. He suggested that when fission resulted in a very unequal partition of mass, the smaller component would be very apt to be a white dwarf.

He concluded this chapter thus: 'We have now sketched in general terms the view of stellar evolution to which we are led by the twin hypotheses of the annihilation of matter as the source of stellar energy and of high ionization as the state of stellar interiors.' But it must be remembered that Jeans's conclusions were based on the theory of the appreciable reduction of mass by radiation, i.e. on the 'long' time-scale of 10^{11} or 10^{12} years, whilst modern opinion tends more and more to the view that the time-scale is of the order of a multiple of 10^9 years, in which case the loss of mass by radiation is negligible. The question, however, is to some extent still open; the dichotomy may possibly find its solution in my own cosmology in which there are two time-scales, one stretching back to a creation 2×10^9 years ago, the other a reinterpretation of this period in dynamical terms which results in its indefinite extension backwards. Some phenomena may have been interpreted by physicists on the one scale, others on the other scale, and until a close analysis is made of the question what is the basic time-scale underlying any given class of phenomena, discrepancies are bound to arise. Jeans's theory of evolution may stand the test of future research better than is supposed at present.

It is a pleasure to turn from these criticisms of Jeans's work on stellar structure to a later physical investigation which he made in the same volume, in which he isolated the phenomenon of radiative viscosity, till then unnoticed by mathematical physicists. In this investigation,* Jeans

* Given in detail in *Astronomy and Cosmogony*, pp. 270–9. The original investigations were published in the *Monthly Notices* of the Royal Astronomical Society in 1926.

showed physical insight of a high order. It had been a commonplace of physics for some years that radiation conveyed momentum—that it exerted a pressure; but it had been overlooked that, in the presence of a gradient of velocity, radiation in conveying momentum from one layer to an adjacent layer would contribute to the apparent action of viscosity in tending to smooth out the velocity gradient. Jeans not only pointed this out, but gave, first, a physical calculation of the coefficient of radiative viscosity based on an analogy with the dynamical theory of gases, in which he was such an expert, and secondly, a more refined calculation of the transfer of momentum in a moving medium. Jeans was thus able to show that in a star the effect of radiative viscosity should be comparable with ordinary gaseous viscosity.

But he made a further interesting application of this effect. If a star is rotating like a rigid body, the flow of radiation from the central regions through the outer regions will have a braking effect on the outer regions, and will tend to slow them down, thereby setting up a departure from rigid-body rotation. Jeans argued out the effect as follows:

> ...the radiation, as it passed further from the axis of rotation, would continually gain in moment of momentum about the axis of rotation [since it would participate in the cross-velocity of the medium]. From the principle of the conservation of angular momentum, this gain must involve a corresponding loss to the layers of the star through which the radiation is passing when the gain occurs. Thus the passage of radiation through a rotating star produces a braking effect on the rotation of the star, and as this effect is different in different layers of the star, it must tend to set up inequalities in the angular velocities of different parts of the star.

He proceeded to embody these ideas in analysis.

I subsequently confirmed Jeans's analysis of the rate of transfer of momentum in a moving medium by a different method, though I arrived at slightly different numerical coefficients for some unexpected terms which arose involving the temperature gradients in the material. All these terms correspond to the fact that the radiation generated at any point carries with it into its surroundings the momentum of the particle from which it took its rise.*

I feel justified in telling the following story respecting Jeans's discovery of the phenomenon of radiative viscosity. The first notification which the scientific world received was the laconic title 'Radiative Viscosity' on the announcement cards which gave the agenda for the meeting of the Royal Astronomical Society at which the papers concerned were to be read. Eddington duly received his card, and at once realized what Jeans's paper was to be about. He sat down and calculated what the coefficient of radiative viscosity should have for its formula, by the very method which Jeans had developed, and sent it on a post-card to Jeans before the meeting. It agreed precisely with the formula Jeans announced at the meeting.†

* See *Monthly Notices* of the Royal Astronomical Society (1929), 89, 518.

† I have consulted the report of the R.A.S. meeting in the *Observatory*: Jeans read his *second* paper on the subject but not his first, and Eddington stated at the meeting that he had looked into the matter independently. But I think that I had the post-card story from Eddington privately.

CHAPTER XII

JEANS AND PHILOSOPHY

In 1942 Jeans published a small volume entitled *Physics and Philosophy*, which I reviewed in *Nature*. In 1947 there was published posthumously a more substantial volume, entitled *The Growth of Physical Science*. (The proofs of this book had been revised by Jeans shortly before his death in September 1946.) The volumes to some extent overlap, since the one specifically dealing with philosophy includes sketches of the history of the progress in physics as the chapter sub-titles indicate: 'The two voices of science and philosophy' (Plato to the present); 'How do we know?' (Descartes to Kant; Eddington); 'The passing of the mechanical age' (Newton to Einstein); 'The new physics' (Planck, Rutherford, Bohr); and 'From appearance to reality' (Bohr, Heisenberg, de Broglie, Schrödinger, Dirac); whilst the volume with the deliberately historical approach has chapters on the remote beginnings of science in Babylonia, Egypt, Phoenicia and Greece; science in Ionia and early Greece (including mathematics, physics, philosophy and astronomy); science in Alexandria, in the Dark Ages, in the Renaissance, in the century of genius (1601–1700), in the two centuries following Newton and in the modern era.

Jeans evidently took great pains over the latter book; it is well illustrated by reproductions of contemporary prints, paintings, etc., and it is indeed a more serious work. *Physics and Philosophy* is of a rather journalistic character, and has been much criticized by professional philosophers.

In my *Nature* review of *Physics and Philosophy* (from which I see no reason to differ today) I quoted Emerson's† definition of philosophy as 'the account which the human

* *Nature, Lond.* (Jan. 16, 1943), 151, 62.

† In his essay on Plato as a 'representative man'.

mind gives to itself of the constitution of the world'. Jeans in his preface modestly disavowed any acquaintance with philosophy other than that of an intruder, and disclaimed any intention to pose as an authority on questions of pure philosophy. Yet, on Emerson's definition of philosophy, no one had a greater right than Jeans to treat the subject of philosophy. Indeed, many of the great mathematical physicists of our generation, starting as mathematical technicians, have been compelled by their own researches to study the philosophy underlying them—Eddington, Planck, Einstein and Schrödinger. Jeans especially, from his studies of the structure and evolution of the universe, was well qualified to have a view on the underlying significance and meaning of it all, on the nature of reality and on free will and determinism. Each generation of men of science has to consider anew the relation of the newly current theories to the perennial problems of philosophy. Though much survives, in each generation, of the scientific achievements of its predecessors, the conclusions of physics, even about itself, have lately changed radically every twenty years or so. This ephemeral character of physical theory contrasts markedly with the changelessness of the grand problems which philosophy presents and always has presented, and necessitates a reconsideration of the answers, not the answers that science *gives* to these questions—for science as such can never answer philosophical problems—but the answers that it can *accept* from philosophers. It should be the business of philosophy to criticize science rather than the business of science to criticize philosophy; but in practice few philosophers become acquainted closely enough with the actual day-to-day tactics of scientific advance, and take even the grand strategy of that advance from its scientific exponents rather than from the content of the advance itself.

From this point of view Jeans's use of philosophy is disappointing. As I have pointed out previously, however critical Jeans was towards the results of his own researches as he conducted them, yet when expounding the present position in physics as it had been developed by other leaders he usually adopted an attitude of acceptance and complacency, attributing tacitly to the conclusions concerning the present position a perfectibility which experience constantly shows not to be justified. For example, Jeans did not himself make contributions to the theory of relativity or the theory of the expansion of the universe. But he never brought his critical powers to bear on the theory of relativity or on the orthodox theory of the expanding universe. He wrote facilely of 'expanding space' in the days when it was the fashion to do so, without ever explaining or even inquiring what in the world this could mean; he even wrote of the nebulae as 'straws showing which way the streams of space were flowing'—a metaphor which, however poetical, is more calculated to darken counsel than to enlighten our minds. I think it was a great pity that Jeans did not live to write the volume of criticism of science that his studies so well equipped him for doing.

On the other hand Jeans was agreeably critical of philosophy as such, though not always self-consistently. Admitting that metaphysics is contiguous to physics, though with a clearly defined boundary, he said that 'the task of physics is to discover and formulate laws, while that of philosophy is to interpret and discuss;* or again, that 'the *métier* of the philosopher is to synthesize and explain facts already known';† but in the former passage he went on to say that 'the physicist can warn the philosopher in advance that no *intelligible* interpretation of the workings of nature is to be expected'.‡ This is flatly contradicted in a later sentence of the same book, that 'knowledge of the external world can come only through observation and experiment. These tell us that the world is *rational*'.§ This

* *Physics and Philosophy*, p. 17. † *Ibid.* p. 96.

‡ *Ibid.* p. 17; *italics* mine. § *Ibid.* p. 82, repeated on p. 174. *italics* mine.

rather cuts away the ground on which he started to be critical of the philosopher's role.

He was on firmer ground in his chapter entitled 'The two voices of science and philosophy'. Here he pointed out that amongst the hindrances to a joint discussion by philosophers and physicists are differences of idiom, if not of language. He stated that, whether one understands the meaning of a sentence in Newton or not, one knows at least the meaning of the words, whereas philosophy has no agreed terminology. He was right in pointing out that various old problems of philosophy owed their existence to imperfections of language; some look very different when translated into the language of science, whilst others vanish away in the process of translation, the different types of problem, for example, involving the word 'infinity', which pure mathematics has removed from the sphere of philosophy. Again he rightly argued that philosophers tend to think of facts as revealed by our primitive senses, whilst science thinks of them as they are revealed by instruments of precision.

He cited Leibniz as trying to invent a calculus for philosophy, in which the symbols would represent a small number of primitive elements or root notions, and by the operation of which disputes between philosophers could be settled. It is not unfair to point out that, before this, John Wilkins, Warden of Wadham and later Master of Trinity, invented a universal language in which the 'characters' stood for ideas; and that the mathematical philosophers, Boole, Russell and Whitehead, did in fact successfully invent a language of the type sought by Leibniz. Jeans did not mention these, but quoted from Anatole France that 'un métaphysicien n'a, pour constituer le système du monde, que le cri perfectionné des singes et des chiens'. Jeans dwelt on this alleged inadequacy of ordinary language to express philosophical ideas or to describe philosophical modes of thought, and criticized Descartes's *cogito, ergo sum* by pointing out that the use of this allowed, for example, no place for telepathy; and he pointed out also how eighteenth- and nineteenth-century physicists, feeling themselves bound to invent a nominative for the verb 'to undulate' (since every verb requires a noun!), thus misled physics for over a century. This is to criticize the use of non-technical language in general, rather than the use of non-technical language by philosophers. I do not think, however, that Jeans profited much by this lesson in language; for in spite of the difficulties in the notion of an ether, to which he was referring, he unashamedly quoted the theory of relativity as giving a structure to space, and even a current flow to space. But Jeans then gave a vivid account of the differences in the meaning of the word 'red' to a physicist and to a philosopher.

As to the further difficulties in the way of an accommodation between physics and philosophy arising from differences of idiom in the two languages used, he argued that the philosopher speaks and thinks in subjective, the scientist in objective, terms; the philosopher in qualitative, the scientist in quantitative, terms. He illustrates by quoting and analysing the age-old situation of two persons differing as to whether the inside of a given room is hot or cold, according to the environment from which they have come. He rightly argued that our senses are not very good at estimating absolute heat and cold: 'We do not judge that an object is hot or cold, so much as that it is *hotter than* or *colder than* something else'.*

Again, he criticized the philosopher's distinction between the so-called primary qualitites which a substance can possess (such as solidity and extension) and its secondary qualities (such as colour or coldness). He pointed out that it is meaningless to imagine a substance being 'stripped' of its so-called secondary qualities, but at the same time necessarily retaining its primary qualities; to the scientist all qualities are primary in the sense that they 'are utterly

* *Physics and Philosophy*, p. 91.

inseparable from the body in what state soever it be'.* But his illustration, that a tulip is not made less red by its being looked at in blue light, was not a very apt one; it merely indicated that the man of science and the philosopher were concentrating on different aspects of the qualities of a substance. There was, however, an authentic Jeans touch in his concluding sentence in this section: 'it is only in Wonderland that a cat can be stripped of everything but a grin'.

He stated further objections to philosophical practice on the ground of its habit 'of depicting the world entirely in black and white, and so ignoring all the half-tones, gradualness and vagueness which figure so prominently in our experience of the actual world'. He instanced the law of the excluded middle 'which has dominated formal logic, with devastating results, from the time of Aristotle on'. 'The scientist...knowing that everything will generally possess some *A*-ness and some not-*A*-ness, is very little concerned as to whether an object is classed as *A* or not-*A*; what he wants to know is how much *A*-ness it possesses.' But again his illustration—an analysis of Zeno's paradox of Achilles and the tortoise—was not really to the point.

He proceeds to criticize St Anselm's so-called ontological argument for the existence of God, and marvels that it imposed itself on later thinkers of the calibre of Descartes and Leibniz, though, as he points out, it did not mislead Kant. He attributes its early success to the false assumption (a consequence of the nature of language) that only two different degrees of existence were recognized, existence and non-existence. But the ontological argument could 'never assign a higher degree of existence to such a Being than existence in our thoughts—*ex nihilo nihil fit*'. Again, 'a man must be either young or not-young'; or he might have instanced the old Greek conundrum, when does a small heap of stones become a large heap? However, I think that

* Quoted by Jeans from Locke.

the verdict on these criticisms by Jeans must be that philosophy does recognize these linguistic difficulties—so much so that some modern philosophers attempt to resolve all philosophical problems into linguistic problems.

Jeans then discussed differences of method between philosophy and science. First, philosophy over-simplified. It did not distinguish between the red of a flower, the red of a sunset, the red of the external galaxies, the red of a fire; or the blueness of the sky and the blueness of an electric fire, the blueness of moonlight and the blueness of shadows on snow. Next, he accused philosophy of being more 'atomistic' than science. It was inclined to see the world as a collection of separate objects, nature as a collection of detached events, and time as a collection of moments. The scientist, he claimed, was more inclined to see 'nature as a theatre of continuous change rather than as a succession of jerks, as a cinematograph show rather than as a series of magic-lantern slides'.* He was again unfortunate in his comparison; for in a cinema exhibition there is only an apparent continuity in the moving pictures, actually it is just what Jeans was accusing the philosopher's view of being, a series of separate snapshots. 'It is the same with space; the philosopher divides this up into small finite regions, but the scientist into infinitesimals....In brief, the philosopher tends to think in terms of what the mathematician calls *finite differences*, whereas the scientist thinks in terms of *infinitesimals*.'† He again drew his example from Zeno, this time the paradox of the arrow in motion, and submitted that every schoolboy could see the fallacy when the matter was expressed in scientific terms; he quoted Berkeley to the effect that infinitesimals are nonsense. But the modern pure mathematician does not believe in infinitesimals either. Jeans's period of education in pure mathematics occurred before the modern era of insistence on arithmetical rigour in laying the foundations of the differential calculus, and I do not

* *Physics and Philosophy*, p. 98. † *Ibid.* p. 99.

think he appreciated that infinitesimals have been abolished in mathematics and replaced by 'differentials'.

He then embarked on a criticism of causality, as expressed by Kant or Bertrand Russell. He says that there is no scientific justification for supposing that the happenings of the world can be divided into detached events, and 'strung in pairs, like a row of dominoes, each being the cause of the event which follows and at the same time the effect of that which precedes'.* He warned his readers at the same time against the other extreme, observing that, in considering any event, it was not necessary for all previous events in the history of the world to be considered as separate causes. For one thing, the effects of the earlier of them were already taken into account in the later, and they need not be allowed for twice over. It was enough to consider the cross-section at one instant of time, and even this cross-section need not extend over the whole of space. Two particular cross-sections, he claimed, were of special interest: first, a cross-section near the beginning of time (the creation of the world); secondly, a cross-section only slightly differing from the present. In the latter case, all those parts of the universe not in our immediate vicinity could be disregarded, and the state of things here and now depended only on the state in our immediate vicinity an infinitesimal moment ago.

Apart from his reactions to traditional philosophy, what were Jeans's views as a whole on ultimate questions? We have seen earlier that his innate reverence for mathematics led him to consider the Creator of the Universe as in essence a mathematician, working with mathematical ideas in a mathematical way. There is no reason to suppose that this was not a genuine belief. But it has not been widely accepted, or developed by others. Whilst objecting to ideas of God as a biologist or an engineer, Jeans still appeared to his contemporaries to be setting undue limits to the nature of God—creating God in his own image, as it were. This

* *Physics and Philosophy*, p. 103.

view was open to the objection that, whilst the idea of God could include the idea of a mathematician, it should not be limited by it; Jeans never writes of God as a supreme, transcendental Being worthy of worship for the creation of the wonderful thing that the universe is. Jeans's 'mysterious universe' is rather a queer collection of insoluble intellectual problems than an edifice of the dim but inspiring vistas that make a Gothic cathedral.

Jeans was a firm believer in relativity as conventionally represented. He adopted the view of Minkowski, that time and space were nothing separately, but merely the resolution by a particular observer of a supposedly genuine physical reality, space-time. He often attacked any realistic view of the existence of an ether, but repeatedly spoke of the theory of relativity as impressing a curvature on the space-time unity, thus letting space-time subserve the purposes for which the nineteenth-century physicists invented the ether. He had a curious habit of referring to the 'ordinary space of physics', distinguishing this in some way from the various spaces presented by the mathematician. He never seems to have faced up to the situation that the choice of a space (within certain limits) in which to depict the phenomena of nature is at the disposal of the theorist: either we can adopt a simple space (Euclidean for example) and let the laws of nature therein found to hold good take up such complications as are necessary; or we can choose simple laws of nature, and put the complications into the space. He wrote of nature 'dodging the problem' of action at a distance by the simple manoeuvre of making gravitation act on space instead of across or through space. He admitted, however, that this only postpones the difficulty; it 'promised a new description, but not a satisfying description of the facts'.

On the subject of free will and determinism, Jeans concluded that the consensus of philosophical opinion today was in favour of determinism of some kind and he himself in numerous instances sought to show that the impression which each man had of the freedom of his own will was illusory. He thought that the existence of indeterminism in the detailed behaviour of electrons and photons, as disclosed by Heisenberg's principle of uncertainty, *might* create a possible formulation of the question in which the verdict would be in favour of free will; but he stipulated that before this could occur, those who believed in the freedom of the will should say exactly what they meant by it—where it differed from unconscious determinism. In one of his similes he wrote:

> The old physics showed us a universe which looked more like a prison than a dwelling-place. The new physics shows us a universe which looks as though it might conceivably form a suitable dwelling-place for free men, and not a mere shelter for brutes—a home in which it may at least be possible for us to mould events to our desires and live lives of endeavour and achievement.

He went on to say that it could hardly be claimed that the new physics justified any new conclusions on determinism, causality or free will, but that the argument for determinism was less compelling than fifty years ago.

Jeans had much to say on the rival merits of materialism and what he called mentalism. He dismissed Berkeley's argument from primary and secondary qualities, on the scientific ground that each quality in question must have an objective part. He dismissed another argument of Berkeley's on the ground that if it was held to be impossible for mind to act on matter, and that hence all was mind, then it could equally be held that all was material, including our mental processes. He quoted Bertrand Russell to the effect that the stuff of the world may be called physical or mental or both or neither, as we please. But it is not clear that Jeans accepted this argument. He surveyed the various contributions which physics might be expected to make to the question. He claimed that the theory of

* *Physics and Philosophy*, p. 216.

relativity had abolished space and time as separate entities, but had been replaced by an objective unity, space-time, which makes space and time neither less real nor more 'mental' than before. He also claimed that relativity had shown that electric and magnetic forces were not real at all, but merely mental constructs of our own, resulting from our attempts to follow the motions of particles; the same was true of gravitation, energy, momentum—all were mental constructs. Only matter itself was left as material.

The new quantum theory took us further away from materialism and towards 'mentalism', with its dual aspect of reality, the wave aspect and the particle aspect. He argued that, 'if we know nothing about a particle except that it exists, all places are equally likely for it', and no number of experiments can locate it exactly. (It is difficult to see what Jeans meant by 'knowing only that a particle existed'; one must have made *some* observations on it to know even this!) He deduced from this that the ingredients of the particle picture were particles existing and moving in physical space, while the ingredients of the wave picture were mental constructs existing and moving in conceptual spaces: the one set were material, the other mental. But he warned his readers that neither picture was a picture of reality—it is merely a picture we draw to help us to imagine the course of events in nature. Hence we are not entitled to regard reality as like the ingredients of either picture. However, he went so far as to say that 'the pictorial representation does not take us into the mansion of reality, but does take us to its doorstep'. By this, he explained, he meant that, if we interpret the wave picture of physical phenomena as a set of waves of knowledge, then reality and knowledge are similar in their natures, that is, reality is wholly mental.

Apart from arguments of the latter type, we could have no means of knowing the true nature of reality. 'The most we can say is that the cumulative evidence of various pieces of probable reasoning makes it seem more and more likely that reality is better described as mental than as material.' It would seem that here Jeans has lost any sense of the difference between scientific reasoning and metaphysical reasoning; he is attempting to use scientific evidence as having a bearing on a metaphysical argument, and swaying a metaphysical conclusion.

But, he continued, admitting that the two entities mind and matter are of the same general nature, which is the more fundamental? For example, are we to suppose, with Berkeley, that the continued existence of any object is due to the existence of some mind *in* which it exists—the mind of God? Or is there an absolute mind comprising all our individual minds? He again seemed to think that an answer to this, the most metaphysical of questions, could be partially answered by physics: the wave picture of the quantum theory—a step 'nearer reality' than the particle-picture—merged all the separate individualities of electrons or photons, much as a raindrop is merged in the sea, and it was conceivable that what was true of perceived objects might also be true of perceiving minds; there might be a wave picture of consciousness, a continuous stream of life beyond the space and time in which individual consciousness existed; beyond space and time we might all be members of one body.

His final verdict on this topic was that the ancient dualism of Descartes, the dualism of mind and matter, no longer contained two antagonistic or mutually exclusive entities, but rather complementary ones; that one no longer needed an elaborate mechanism to keep them in step; but that the one controlled the other; just as in physics, waves control particles, so the mental controls the material. In putting this as a simile, I may be misrepresenting Jeans. I am not sure that he did not really believe that the realities represented by particles and waves were identical with what earlier philosophers called just matter and mind.

This seems a convenient place to criticize again Jeans's views on astronomical eschatology—the end of the universe. Jeans never lost an opportunity of representing the universe as tending to what is called a 'heat-death', namely a state in which all differences of temperature, chemical composition, state of aggregation, etc., are evened out, in which none of the energy in the universe is capable of being used to conduct processes but all sources of available energy have been merged into the corresponding sinks. This conclusion is reached by an uncritical application of the second law of thermodynamics, according to which the entropy of any finite isolated system tends to a maximum, a dull featureless uniformity. The supposed process has been compared to the running down of a clock. It invites the obvious comment that for a clock to run down, it must first have been wound up; if so, who wound it up, and cannot the great Winder-up of the universe wind it up again if it runs down? Jeans claimed to find no hope of escape from this pessimistic conclusion of a heat-death in physical science. The second law of thermodynamics was inexorable, and permitted of no reversal of the sense of direction in which the universe was moving.

But scrutiny of the details of the so-called proof, that entropy must always increase, shows that the proof is only valid in well-defined circumstances. It applies only to a finite portion of the universe; and it involves the condition that even such a finite portion can be divided into two portions, phenomena in one of which can be regarded as not directly affecting phenomena in the other, and one of which can be regarded as a sort of cosmical private laboratory in which reversible changes of entropy can so take place as to afford a means of measuring the entropy-increase equivalent of any irreversible process occurring in the other. It is only by devising a reversible process (along which entropy changes are measurable) capable of reversing a so-called irreversible change in the one part of the universe that the latter change can be shown to be associated with an increase of entropy. This is not the place to go into technical details. But it should have been clear to all philosophers and theologians (many of whom took too easily on trust the dogmatic assertions of influential physicists) that so tremendous a theorem as the heat-death theorem could only be established under very special and indeed artificial conditions. Such conditions do not hold in the universe considered as a whole. In the first place the universe may be infinite in the amount of matter it contains; in the second place it is expanding and not stationary, and an attempt to supply a rigorous proof of the heat-death theorem for a steadily expanding system would meet with great difficulties; in the third place it is not possible to imagine the universe divided into two portions such that all processes in one portion leave the state of the second portion unaltered. All these considerations give reason for pausing before sentencing the universe to a heat-death. The most that could be said would be that our own galaxy may eventually suffer a heat-death. But there are countless other galaxies in the universe, and the overwhelming majority of these (according to kinematic relativity) are at the present moment experiencing conditions arbitrarily close to those of creation. Our own galaxy is the oldest, but for one galaxy that degenerates to a heat-death there will be a host that have only just started on their careers, and beyond them again arbitrarily many that are as young as we care to specify.

Jeans's own studies in the realm of the second law of thermodynamics were all concerned with the kinetic theory of gases, in which the specimen under discussion is supposed walled around in a finite vessel; and to such systems the notion of a heat-death is applicable. But by no means is the same result to be predicted of the whole universe. I consider that Jeans held this view too dogmatically. I once wrote a short paper embodying the above ideas (in greater technical detail) and sent it to the Royal Society, during

Jeans's secretaryship. It was duly rejected. I had at one time the intention of expressing these views at the discussion on the evolution of the universe held by the British Association at its Centenary Meeting in London in 1931. This discussion was to be opened by Jeans. I happened to meet him (I think at an afternoon 'at-home' at the National Physical Laboratory) some time before the discussion, and mentioned my intention, stating that I had once communicated a paper on the subject to the Royal Society. Jeans frowned, said that the paper had passed through his hands as Secretary (there must have been an independent hostile referee as well), that he thoroughly disagreed with its conclusions, and that he would prefer that I should not bring this topic into what I intended to say at the B.A. discussion. I felt that, under these circumstances, to dispute the general view of the heat-death of the universe on rather technical grounds at a general discussion would be out of place; and I spoke on another topic. I had also a genuine suspicion that I might myself be technically wrong, and I deferred to Jeans's great authority. That was before my own researches in cosmology. But I am now convinced that an unconditional prediction of a heat-death for the universe is an over-statement.

BIBLIOGRAPHY

Technical Books

1904 *The Dynamical Theory of Gases.* 2nd ed. 1916, 436 pp. Cambridge Univ. Press.
1906 *Theoretical Mechanics.* Boston. Ginn.
1908 *The Mathematical Theory of Electricity and Magnetism.* 2nd ed. 1911, 584 pp. Cambridge Univ. Press.
1914 *Report on Radiation and the Quantum Theory.* Physical Society Report, 90 pp. London. 'The Electrician' Publishing Co.
1919 *Problems of Cosmogony and Stellar Dynamics.* Adams Prize Essay, 1917, 293 pp. Cambridge Univ. Press.
1923 *The Nebular Hypothesis and Modern Cosmogony.* Halley Lecture, 1922, 31 pp. Oxford: Clarendon Press.
1928 *Astronomy and Cosmogony,* 420 pp.; 2nd ed. 1929, 428 pp. Cambridge Univ. Press.
1940 *Introduction to the Kinetic Theory of Gases,* 311 pp. Cambridge Univ. Press.

Popular Books

1926 *Atomicity and Quanta.* Rouse Ball Lecture, 1925, 64 pp. Cambridge Univ. Press.
1929 *The Universe Around Us,* 352 pp. Cambridge Univ. Press.
1930 *Eos, or the Wider Aspects of Cosmogony,* 88 pp. London: Kegan Paul.
1930 *The Mysterious Universe.* Rede Lecture, 1930, 154 pp. Cambridge Univ. Press.
1931 *The Stars in their Courses,* 200 pp. Cambridge Univ. Press.
1933 *The New Background of Science,* 303 pp. Cambridge Univ. Press.
1934 *Through Space and Time,* 224 pp. Cambridge Univ. Press.
1938 *Science and Music,* 258 pp. Cambridge Univ. Press.
1942 *Physics and Philosophy,* 222 pp. Cambridge Univ. Press.
1947 *The Growth of Physical Science,* 364 pp. Cambridge Univ. Press.

Original Papers

1900 The striated electrical discharge. *Phil. Mag.* 49, 245–62 and 1, 521–9.
1901 The distribution of molecular energy. *Phil. Trans.* A, 196, 397–430.
1901 The theoretical evaluation of γ. *Phil. Mag.* 2, 638–51.
1901 The mechanism of radiation. *Phil. Mag.* 2, 421–55.

Bibliography

1902 The equilibrium of rotating liquid cylinders. *Phil. Trans.* A, **200**, 67–104.

1902 Conditions necessary for equipartition of energy. *Phil. Mag.* **4**, 585–96.

1902 The stability of a spiral nebula. *Phil. Trans.* A, **199**, 1–53.

1903 The vibrations and stability of a gravitating planet. *Phil. Trans.* A, **201**, 157–84.

1903 The kinetic theory of gases developed from a new standpoint. *Phil. Mag.* **5**, 597–620.

1903 On the vibrations set up by molecular collisions. *Phil. Mag.* **6**, 279–86.

1904 The kinetic theory of gases. *Phil. Mag.* **6**, 720–2 and **7**, 468–9.

1904 A suggested explanation of radioactivity. *Nature, Lond.*, **70**, 101.

1904 The determination of the size of molecules from the kinetic theory of gases. *Phil. Mag.* **8**, 692–9.

1904 The persistence of molecular velocities in the kinetic theory of gases. *Phil. Mag.* **8**, 700–3.

1905 Gas-theory and radiation. *Nature, Lond.*, **72**, 101–2.

1905 The partition of energy between matter and ether. *Phil. Mag.* **10**, 91–8.

1905 Radiation theories compared. *Nature, Lond.*, **72**, 293–4.

1905 Statistical mechanics applied to ether and matter. *Proc. Roy. Soc.* A, **76**, 296–311.

1905 The density of algol variables. *Astrophys. J.* **22**, 93–102.

1905 The law of radiation. *Proc. Roy. Soc.* A, **76**, 545–52.

1906 The constitution of the atom. *Phil. Mag.* **11**, 604–7.

1906 The thermodynamic theory of radiation. *Phil. Mag.* **12**, 57–60.

1906 The H-theorem and the dynamical theory of gases. *Phil. Mag.* **12**, 80–2.

1906 The stability of submarines. *Nature, Lond.*, **74**, 270.

1906 The deduction of Wien's law. *Phys. Z.* **7**, 667 and **8**, 91–2.

1908 Radiation theory. *Phys. Z.* **8**, 853–5.

1909 The temperature of radiation and the partition of energy. *Phil. Mag.* **17**, 229–54.

1909 The motion of electrons in solids. Part I. *Phil. Mag.* **17**, 773–94 and Part II, **18**, 204–26.

1910 The motion of a particle about a doublet. *Phil. Mag.* **20**, 380–2.

1910 The radiation from electronic orbits. *Phil. Mag.* **20**, 642–51.

1910 Planck's radiation theory and non-Newtonian mechanics. *Phil. Mag.* **20**, 943–54.

1913 The kinetic theory of star clusters. *Mon. Not. R. Astr. Soc.* **74**, 12.

1914 Gravitational instability and the nebular hypothesis. *Phil. Trans.* A, **213**, 457–85.

Bibliography

1914 Radiation and free electrons. *Phil. Mag.* **27**, 14–22.

1915 The potential of ellipsoidal bodies and the figures of equilibrium of rotating liquids. *Phil. Trans.* A, **215**, 27–78.

1915 On the theory of star-streaming and the structure of the universe. *Mon. Not. R. Astr. Soc.* **76**, 70–84.

1916 The law of distribution in star clusters. *Mon. Not. R. Astr. Soc.* **76**, 567–72.

1916 The theory of star-streaming. *Mon. Not. R. Astr. Soc.* **76**, 552–67.

1916 The instability of the pear-shaped figure of equilibrium. *Phil. Trans.* A, **217**, 1–34.

1917 Rotation as a factor in cosmic evolution. *Mon. Not. R. Astr. Soc.* **77**, 186–99.

1917 Gravitational instability and the figure of the earth. *Proc. Roy. Soc.* A, **93**, 413–17.

1917 Internal motions in spiral nebulae. *Observatory*, **40**, 60–1.

1917 The radiation of the stars. *Nature, Lond.*, **99**, 365.

1917 Cosmogonic theories and the motion of tidally distorted masses. *Mem. R. Astr. Soc.* **62**, 1–45.

1917 Note on the action of viscosity in gaseous and nebular masses. *Mon. Not. R. Astr. Soc.* **77**, 200–4.

1917 The equations of radiative transfer of energy. *Mon. Not. R. Astr. Soc.* **78**, 28–36.

1917 The evolution and radiation of gaseous stars. *Mon. Not. R. Astr. Soc.* **78**, 36–47.

1918 The present position of the nebular hypothesis. *Scientia, Bologna*, **24**, 270–81.

1918 The evolution of binary systems. *Mon. Not. R. Astr. Soc.* **79**, 100–6.

1919 (Bakerian lecture.) The configurations of rotating compressible masses. *Phil. Trans.* A, **218**, 157–210.

1919 The internal constitution and radiation of gaseous stars. *Mon. Not. R. Astr. Soc.* **79**, 319–32.

1919 The origin of binary systems. *Mon. Not. R. Astr. Soc.* **79**, 408–16.

1922 The motion of stars in a Kapteyn Universe. *Mon. Not. R. Astr. Soc.* **82**, 122–32.

1922 The dynamics of moving clusters. *Mon. Not. R. Astr. Soc.* **82**, 132–9.

1923 Theory of the scattering of α and β rays. *Proc. Roy. Soc.* A, **102**, 437–53.

1923 The propagation of earthquake waves. *Proc. Roy. Soc.* A, **102**, 554–74.

1923 On the tidal distortion of rotating nebulae. *Mon. Not. R. Astr. Soc.* **83**, 453–58.

1923 The mechanism and structure of planetary nebulae. *Mon. Not. R. Astr. Soc.* **83**, 481–93. Supplement 1923.

1923 Internal motions in spiral nebulae. *Mon. Not. R. Astr. Soc.* **84**, 60–76.
1924 Cosmogonic problems associated with a secular decrease of mass. *Mon. Not. R. Astr. Soc.* **85**, 2–11.
1925 On the masses, luminosities and surface temperatures of the stars. *Mon. Not. R. Astr. Soc.* **85**, 196–211.
1925 On the masses, luminosities and surface temperatures of the stars. Second paper. *Mon. Not. R. Astr. Soc.* **85**, 394–403.
1925 On the masses, luminosities and surface temperatures of the stars. Final note. *Mon. Not. R. Astr. Soc.* **85**, 792–7.
1925 On a theorem of von Zeipel on radiative equilibrium. *Mon. Not. R. Astr. Soc.* **85**, 526–30 and Supplement 1925, **85**, 933–5.
1925 Note on the distance and structure of the spiral nebulae. *Mon. Not. R. Astr. Soc.* **85**, 531–4.
1925 On Cepheid and long period variation and the formation of binary stars by fission. *Mon. Not. R. Astr. Soc.* **85**, 797–813.
1925 The effect of varying mass on a binary system. *Mon. Not. R. Astr. Soc.* **85**, 912–14.
1925 A theory of stellar evolution. *Mon. Not. R. Astr. Soc.* **85**, 914–33.
1926 The radiation from a pulsating star and from a star in process of fission. *Mot. Not. R. Astr. Soc.* **86**, 85–93.
1926 On radiative viscosity and the rotation of astronomical masses. *Mon. Not. R. Astr. Soc.* **86**, 328–35. 2nd paper, 444–58.
1926 Stellar opacity and the atomic weight of stellar matter. *Mon. Not. R. Astr. Soc.* **86**, 561–74.
1926 The exact equation of radiative equilibrium. *Mon. Not. R. Astr. Soc.* **86**, 574–8.
1926 Note on the internal densities and temperatures of the stars. *Mon. Not. R. Astr. Soc.* **87**, 36–43.
1927 On liquid stars and the liberation of stellar energy. *Mon. Not. R. Astr. Soc.* **87**, 400–14.
1927 On liquid stars—configurations of stability, long-period variables and stellar evolution. *Mon. Not. R. Astr. Soc.* **87**, 720–39. Supplement 1927.
1928 Liquid stars, a correction. *Mon. Not. R. Astr. Soc.* **88**, 393–5.

Lectures

The following are references to abstracts of lectures:

1923 The radiation problem. (8th Guthrie lecture.) *Proc. Phys. Soc., Lond.*, **35**, 222–4.
1923 The physical significance of van der Waal's equation. (Van der Waals memorial lecture.) *J. Chem. Soc.* **123**, 3398–414.
1925 Electric forces and quanta. (16th Kelvin lecture.) *Nature, Lond.*, **115**, 361–8.
1928 The wider aspects of cosmogony. *Nature, Lond.*, **121**, 463–70.

Bibliography

1928 The physics of the universe. (H. H. Wills memorial lecture, Bristol.) *Nature, Lond.*, **122**, 689–700.

1931 The annihilation of matter. *Nature, Lond.*, **128**, 103–10.

1931 Beyond the Milky Way. *Nature, Lond.*, **128**, 825–32.

1931 The origin of the solar system. *J. Franklin Inst.* **212**, 135–45 and *Nature, Lond.*, **128**, 432–435.

1932 What is radiation? (14th Sylvanus Thompson memorial lecture.) *Brit. J. Radiol.* **5**, 21–37.

1936 The size and age of the universe. *Not. Proc. Roy. Instn.* **29**, 1, 65–86 and *Nature, Lond.*, **137**, 17–24.

1945 The astronomical horizon. (Deneke lecture, 1944.) 23 pp. Oxford: Clarendon Press.

INDEX

Index

Index

Index

Index